Serendipity, Luck and Wisdom in Research

Serendipity, Luck and Wisdom in Research

Patrick J. Hannan

iUniverse, Inc.
New York Lincoln Shanghai

Serendipity, Luck and Wisdom in Research

iUniverse books may be ordered through booksellers or by contacting:

iUniverse
2021 Pine Lake Road, Suite 100
Lincoln, NE 68512
www.iuniverse.com
1-800-Authors (1-800-288-4677)

ISBN-13: 978-0-595-36551-7 (pbk)
ISBN-13: 978-0-595-80982-0 (ebk)
ISBN-10: 0-595-36551-5 (pbk)
ISBN-10: 0-595-80982-0 (ebk)

Printed in the United States of America

My four children, Cathy, Peg, Marilyn, and Tom, have endured years of discussions concerning the appeal of potential serendipitous tales for this book. Their patience has been commendable, and to them I dedicate this work.

Contents

Foreword.......xiii

Chapter 1: Serendipity: How It Happens....... 1

Chapter 2: The Advantages of Being Lucky and Observant in Chemistry Laboratories....... 27

Chapter 3: Soaring, Bubbling, Lightning—Where Air Meets Sea....... 50

Chapter 4: Penicillin: Some Little-Known Factors....... 60

Chapter 5: An Improbable Route to a Cure for Tuberculosis....... 68

Chapter 6: Contributions of Aspirin, Rat Poisons, and Other Compounds to Health....... 78

Chapter 7: Soil Microbes, Anesthetics, Sunken Submarines, and Other Biological Tools....... 96

Chapter 8: Learning from Animals and Insects....... 115

Chapter 9: From the Earth to the Stars....... 124

Chapter 10: Winning the Big One....... 141

Chapter 11: Technical Literature: Pluses and Minuses....... 155

Chapter 12: Chance and Finding One's Niche....... 165

Chapter 13: Keeping the Monkey Wrench Out of the Gears or How Not to Screw Up Research....... 181

Chapter 14: A Philosophy for Science 208
Index 217

Acknowledgments

The list of friends and fellow scientists who have contributed to this book is long and undoubtedly incomplete, because the time between my original interest in serendipity and the date of publication is more than forty-five years. I am most grateful to my fellow organizers of the 1986 American Association for the Advancement of Science symposium, Rustum Roy and the late John Christman, for their interest and inspiration. Without the cooperation of Jean'ne Shreve, who was chairperson-elect of the chemistry section of AAAS, the symposium could not have taken place. I am most grateful to her and to those who made presentations at the symposium: Roy Plunkett, Lawrence David, Kurt Nassau, Ann Clamandon, and James Moran. Sincere thanks are due my good friend and collaborator, Damian Jones, whose continued encouragement and help in many incidental tasks was invaluable.

Part of the impetus for writing this book was the publication of four articles in *Chemtech* describing much of the material covered in the 1986 symposium. Initially one article was submitted but, with the encouragement and assistance provided by Dorit Noether and the late Ben Luberoff, the coverage was greatly extended. I am grateful for their support.

Following the first draft of the book, I was helped greatly by the late Celeste Kirk. Her vast experience helped to provide organization for what consisted of relatively disparate singular instances of serendipity.

To search through years of *Current Contents* issues would require filling out countless request slips at the National Medical Library on the grounds of the National Institutes of Health. I was spared that tedious task by the understanding research staff at that wonderful library. Similarly, I was given access to the stacks at the National Agricultural Library at Beltsville, Maryland, so that I could leaf through many volumes bearing the title of *Annual Review of* ——. Circumstances controlling such permissions would make privileges such as this almost impossible now.

My admiration of the Naval Research Laboratory where I worked for many years is evident in chapter 13 of this book. In fact, the decision to include the word "wisdom" in the title of the book was dictated by the philosophy I came to appreciate at NRL. Research is defined, in part, as "investigation or experimentation aimed at the discovery and interpretation of facts"(Webster's New Collegiate Dictionary, 1979). No one can argue with that definition, but experience has shown that major advances have also been made as the result of observations that were far removed from an orderly progression of thought. How does a scientist who is indoctrinated with the scientific method reconcile the difference between a rigorously planned study and an incidental, fortuitous fact that seems promising? For many years I was privileged to have as my supervisor John Leonard, who had a broad understanding of science. John was rigorously logical in his interpretation of research results, but, at the same time appreciated the possibility that unforeseen developments might merit closer study. The same can be said of Homer Carhart, whose accomplishments are featured in that chapter. I am most grateful that I knew both of these men so well. Another extremely helpful accomplice at NRL was Kathy Parrish, who took the time to dig out of the files many of the pictures that are included in this book.

To all of the respondents from the inquiries sent to members of the Royal Society and the National Academy of Sciences, I am very grateful. It was a pleasant surprise to receive responses from so many renowned individuals who took the time to respond to the questions I asked of them: J. C. Skou, Carl Woese, Robert M. Berne, Eric C. Conn, Sir David Cox, Ross E. Davis, Robert E. Dickinson, O. H. Frankel, Robert C. Gallo, W. Thirring, Henry M. Hoenigswald, J. Woodland Hastings, Frank Hahn, Herbert Hauptman, T. Kent Kirk, Alfred G. Knudson, Jr., Robert J. Lefkowitz, Donald H. Levy, Daniel McKenzie, Robert H. Purcell, Edwin Roedder, Heini Rohrer, George N. Somero, Endel Tulving, Sir Richard Southwood, George D. Watkins, M. Gordon Wolman, Morris Halle, John H. Crowe, Walter Bodmer, Brigitte A. Askonas, Mark S. Bretscher, Allan S. Hay, Martin Aitken, G. J. Van Nossal, M. H. P. Bott, J. C. Polanyi, R. M. Laws, Eugene Roberts, Thomas F. Anderson, Baruch S. Blumberg, Thomas B. Roberts, Donald Walker, John M. Edmond, W. Reichardt, A. D. Bangham, Michael J. Berridge, G. Burnstock, Fergus W. Campbell, Dennis Chapman, Geoffrey Eglinton, Noreen E. Murray, Edward C. Cocking, A. Dalgarno, L. G. Goodwin, Robin Holliday, Edward Irving, and K. L. Johnson.

Throughout much of my career at the Naval Research Laboratory, I was a member of the Thomas Edison Toastmasters Club. Typically such a club provides critical reviews of the speeches being presented, and the lessons learned there have carried over into the writing of this book. I am particularly mindful of the interest shown by Rex Neihof, Kathy Parrish, and Tom Gordon. In recent years there have been many lunches shared with Joe Aviles, Joe Dytrt, and Perry Alers, in which we discussed serendipity along with national issues of slightly less importance.

At many times I have relied on others for editorial help, including Dale Bultman, Kingsley Williams, Al Herner, Robert Lamontagne, Bill Barger, Wally Brundage, and John Leonard. I am particularly grateful to my daughter, Peg Laramie, whose long experience in journalism was most helpful in appraising the wealth of material that I had uncovered through the years.

In my postretirement years I have benefited from having many friends at the Environmental Fate and Effects Division, in the Office of Pesticide Programs at the U.S. Environmental Protection Agency, Crystal City, Virginia. No one has a greater appreciation of serendipity than Alex Clem, whose enthusiasm for the subject has been treasured. Others who have proofread much of what appears in this book are Susan Lees, Dana Spatz, Tom Kopp, Faruque Khan, Ron Harper, Dave Jones, Kevin Costello, Leo LaSota, Tom Steeger, Nelson Thurman, Jim Hetrick, Ed Fite, Ed Odenkirchen, Brian Montague, Minh-Thuy Nguyen, Jim Carlton, Silvia Termes, and Skee Jones. For solving innumerable computer problems for me, I will always be grateful to Tom Kopp.

Finally, I express extreme gratitude to Cathy Hotka and Peg Laramie, my daughters, whose grasp of technical details concerning the art of writing has been completely invaluable.

Foreword

A common perception of scientists is that they work in an ivory tower on subjects too technical for the average person to comprehend. That is often the case, but there have been many instances in which fortuitous circumstances led to discoveries more valuable than the goal so carefully sought. When researchers begin laboratory studies, there is often no way of knowing the obstacles that will have to be overcome or whether time is on their side. Furthermore, even a successful experiment might produce a huge payoff in another field years later. A good example is a serendipitous finding made in London in 1946 that was important in its own right but considerably more valuable in another field. The goal of that initial research was to develop a procedure for preserving the sperm of fowl. About twenty-five years later the technique became an essential component in the replacement of corneas in the human eye.

There were several sources used for material in this book, including responses received from members of the National Academy of Science in the United States and the British Royal Society. I wrote to approximately sixty members of each group and received responses from half of them. Two questions were asked: "Were there any unplanned circumstances that directed you into the field in which you made your mark as a scientist?" and "Did chance play a significant role in your accomplishments?" The replies ranged from short but polite negative statements to fascinating narratives several pages long.

Second, the periodical *Current Contents* carried a feature for about fifteen years entitled *Citation Classics* in which authors of frequently-cited research papers would give informal accounts of the circumstances surrounding those studies. In many, authors of these important papers told of their difficulty in getting them published, because the reviewers underestimated their value. In many others were the accidental discoveries that made the results possible.

A third delightful mine of information was found in prefatory chapters of publications whose titles begin with *Annual Review of*——. There are more than a hundred such annual reviews covering a wide range of science. These were written by senior scientists who reveled in relating the actual events that led to their most important discoveries. Included were instances of hunches that paid off, or simply serendipitous happenings that could not have been foreseen.

Last, much inspiration for this effort stemmed from a symposium on serendipity that took place in 1986 as part of the American Association for the Advancement of Science (AAAS) meeting in Philadelphia, Pennsylvania. This was organized by Rustum Roy from Penn State University, the late John Christman from Loyola University in New Orleans, and myself, from the Naval Research Laboratory, Washington, DC. Prior to the symposium we had solicited input from readers of approximately forty-five scientific journals. There were many responses to this appeal and the greatest input came from readers of the British journal *Nature.* The night before this symposium there was a cocktail party for the speakers (who had not met each other), and the informal atmosphere thus generated carried over into the symposium the next day.

It has been a pleasure to write this book, which should appeal to all who enjoy a good story, regardless of background. My formal training was in chemistry; I received a bachelor of science degree in that subject from Catholic University (Washington, DC) in 1942, then a master of science degree in organic chemistry from the same institution in 1948. I spent the World War II years at the Geophysical Laboratory of the Carnegie Institution of Washington, where research was being conducted on improving the linings of gun barrels. During that time I was privileged to be one of the early users of carbon-14 (^{14}C) as a tracer, which was used to determine the penetration of carbon into those linings. My career with the U.S. government began in 1948 with the Insecticide Division of the Department of Agriculture Laboratories in Beltsville, Maryland. This was followed by seven years at the Engineering Research and Development Laboratories at Fort Belvoir, Virginia, where fungicides were my principal interest. I moved to the Naval Research Laboratory in 1956 and retired from that distinguished organization in 1987. While at NRL one of my interesting tasks concerned the problem of maintaining a habitable atmosphere in nuclear submarines. This required that the carbon dioxide exhaled by the men be removed, and replaced by oxygen. Through the process of photosynthesis, mass cultures of microscopic algae should accomplish that task. Results of that laboratory study were far different than I had anticipated, in that the system was extremely reliable in removing carbon dioxide and producing oxygen in a flowing air stream. The problem was that the electrical power required for the incandescent lights necessary for photosynthesis was prohibitive. Following those studies, I pioneered in devising bioassays based on the gas exchange properties of microorganisms. With such an approach, I showed that the response times of microorganisms is orders of magnitude faster than bioassays that typically depend on measurements of mass or cell numbers. I was able to detect the toxicity of stainless steel to algae in less than fifteen minutes. Since 1989 I have been a part-time employee of the American Association of Retired Persons, later associated with the National Older Workers Career Center, and my assignment is at the Crystal City, Virginia, offices of the Environmental Protection Agency.

Chapter 1

Serendipity: How It Happens

"One of the reasons I think history is endlessly interesting is that nothing ever had to happen the way it happened. It was never on a track, never preordained."

—David McCullough

The Power of Positive Thinking

The medical community acknowledges that someone with a strong will to live has a better chance of overcoming a life-threatening illness than someone not so predisposed. Will-to-live motivations can be prompted by personal goals, such as achieving a long-standing objective, but less tangible factors can also be important. One of these is a sense of humor.

A humorous person, almost by definition, must be creative, but the converse need not be true. Nurturing a sense of humor, playfulness, revelry, or the like is not included in the technical training of scientists, but it can play a necessary role in reaching an objective. A compelling example is found in the life of Norman Cousins, the former editor of the *Saturday Review.* He had a painful arthritic condition, and when doctors told him that his condition was degenerative and there was little they could do except treat his symptoms, he took matters into his own hands. He devised a regimen of laughter built around old comedies, such as those with the Marx brothers, or reruns of *Candid Camera.* For several hours a day he enjoyed belly laughs and to his surprise and the disbelief of the doctors, he recovered from the arthritic condition that had threatened to enslave him. Cousins wrote a book on the subject and gained enough credibility to be invited to join the faculty of the medical school at the University of California at Los Angeles (UCLA) where his ideas on the beneficial aspects of laughter were tested scientifically and shown to be valid. Subsequently he wrote a book[1] describing his role in the development of UCLA's program in psychoneuroimmunology. Humor was a prominent aspect of this *Biology of Hope,* as was the power of mental concentration.

One of the stories he told concerned the success he had in increasing the rate of blood flow to his hands to foster a recovery from an injury simply by concentrating his mind on the subject. Later, he suffered a fracture of the radial bone in his elbow as the result of a fall on the tennis court. With his arm in a cast and sling, he attended a concert at the Hollywood Bowl and by chance was in the box next to Dr. Frank Jobe. Athletes know of orthopedist Jobe because of his reputation in creative reconstructive surgery. Jobe inquired of the evident problem that Cousins had and cautioned Cousins that such an injury would take months to heal because of the restricted blood supply to the elbow. Upon reflection, Cousins reasoned that if he could increase the blood flow to his hands, then he could surely do the same for his elbow. After three or four attempts to influence the blood flow, he noticed increased movement in his fingers. He repeated the exercise several times each day for the next week with progressively beneficial results. About ten days later he took off the cast and sling. A week after that he was back on the tennis court.

Cousins's experiences at UCLA over a ten-year period support his claim that the state of mind plays a large role in one's health. Further, he provides an insight on how to judge a man. It is taken from the writings of Dostoyevsky:

> If you wish to glimpse inside a human soul and get to know a man, don't bother analyzing his ways of being silent, of talking, of weeping, or seeing how much he is moved by noble ideas; you'll get better results if you just watch him laugh. If he laughs well, he's a good man.

Long Odds on the High Seas

A black and stormy night at sea is not conducive to success in a research project, but its effects might still be overcome by luck. This is from a long account supplied by David C. Nobes, Department of Physics, University of Waterloo, Ontario, Canada, who responded to the plea that we sent to many journals asking their readers for instances of serendipity. His team was studying the electrical structure of the seafloor using marine electromagnetic sounding techniques. Part of the methodology required the use of an ocean bottom magnetometer (OBM) that had two perpendicular channels.

On any cruise the researchers would put the OBMs down at a number of sites, and on this particular cruise there were three sites at which they dropped two OBMs each. On the third drop, OBM #1 was placed in an area of very high heat flow. The night gradually became more and more stormy. At about 1:00 AM the researchers got a call from the bridge that a flashing strobe light had been spotted (each OBM had both a strobe light and a radio beacon). Sure enough, there was a radio signal, and it was coming from one of their OBMs, even though they had not planned for any of them to pop up at that time. They spent the next two hours hauling in their transmitter and then went after the OBM which was drifting north/northwest at a furious rate, perhaps destined to wash up on some remote Alaskan shore. They finally recovered it—it was OBM #1. When the data was dumped from it, they learned that it had been recording data until midnight. Fortunately, it had popped up early and drifted across their bow just as they were turning to head southeast. They were lucky to detect it, because it would have been lost had it not drifted by their bow long before they had planned on being where they were. The scientific merit of finding this OBM was that it had been on a hydrothermal vent field. The vast majority of such vents are on hard rock, but this one was on top of five hundred meters of sediments. This was a surprise, but Nobes said the area was under study anyway and would have been found regardless. Still, it was a nice sideline.

There is nothing more black than a night at sea when clouds dominate the sky; to find anything under those circumstances defies all odds. Luck was the key factor.

One Thunderstorm That Had a Payoff

My good friend, Dave Flemer, while working at the University of Maryland's Chesapeake Biological Laboratory, was coauthor of a paper titled *Power Plants: Effects of Chlorination on Estuarine Primary Production.* The scene of this study was the steam electric station at Chalk Point, Maryland, on the Potomac River estuary. The purpose was to explore the environmental effects of the temperature rise in the water that was circulated through the heating coils at the station, which involved photosynthetic rates based on measurements of radioactively tagged carbon, ^{14}C. Samples taken from the inlet of the plant were compared with those taken from the outlet fifteen minutes later, because it had been determined that it was the time required to transit the plant. One day a substantial thunderstorm was about to hit the area, and Flemer didn't relish the idea of working outside in the open. He obtained permission from the plant operator to move his incubation chambers inside a nearby metal shed until the storm subsided. In this shed were the controls for periodically injecting chlorine into the water to reduce the rate of fouling of the heat exchangers. While attending the samples, he noticed the significant odor of chlorine because the solenoids leaked somewhat. Of course, Flemer was alerted to the possibility that the chlorine might affect the productivity measurements he was making. When he studied the data he had obtained to determine the effect of the water being heated, there was little effect, but a comparison of samples taken over various time periods showed large variations, and these coincided with the times that chlorine had been injected. The figures showed that there could be as much as a 6.5 percent reduction of primary productivity for the affected part of the river.

Work done by the group at nearby Chesapeake Biological Laboratory at Solomons, Maryland, also benefited from an astute observation made by a young student. The subject was the decline of grasses growing in the Chesapeake Bay, and one of the potential causes of this phenomenon was the runoff of herbicides such as atrazine and diuron. Among those who made the trip to the bay to assess the situation was the student, who had experience working on a farm. When he looked at the grasses in the bay, he was impressed with the "fuzziness" of the plants because there was a resemblance to what he had seen in the ponds on his farm. He concluded that this could well be due to the runoff from fertilizers and pesticides from farms because that was the situation he had experienced. At the time (late 1960s) there had not been as much awareness of the nutrient runoff as there is now.

A Slip of the Tongue

When the tongue says what the subconscious mind tells it, we call it a slip. Everyone has experienced such an event, but the result is normally too trivial to think about. In the case of a Nobel Prize winner, it might be more important.

Murray Gell-Mann won the 1969 Nobel Prize in physics because of his work on the theory of elementary particles. Whether the event described here played a role in that prize is not mentioned in the source from which it came, but it surely was an accident that had a payoff.[2] He was lecturing about strange atomic particles. The word "strange" was chosen because while they are produced in great numbers by strong forces, they have a long lifetime and decay slowly through weak interactions. The conventional wisdom endorsed by the scientific community had been that the values of isotopic spin had to be half-integer values, such as 3/2 or 1/2. Only an atomic physicist would understand exactly Gell-Mann's slip of the tongue, but the message Gell-Mann gave was clear:

> One day I was giving a talk at Princeton on why conservation of isotopic spin failed to work as an explanation of the long lifetime and, mentioning the {baryons}, I was going to say, "Suppose they have (I) = 5/2," but said "(I) = 1" instead. I stopped dead because I realized that (I) = 1 would do the job. I saw immediately that the requirement that {baryons} have these 1/2 integral values was just a superstition.

The Miracle of Teflon

The foreword of this book mentions the value of the symposium on serendipity as part of the 1986 AAAS meeting in Philadelphia. An unforgettable aspect of that meeting was the opportunity it provided to talk with Roy Plunkett, who invented Teflon. At his own expense he and his wife, Lois, came to Philadelphia from Corpus Christi, Texas, and they were the hit of the cocktail party in my room the night before the symposium. Roy Plunkett was a friendly, reserved person, totally devoid of any pretense of grandeur, though his discovery of Teflon was surely one of the more memorable chemical events of the twentieth century.

His story began with a problem associated with the early (1928) mechanical refrigerators. They used either sulfur dioxide or ammonia as a refrigerant gas, and either type worked well; when they sprung a leak, however, there was great anxiety in the kitchen. Sulfur dioxide provokes a choking reaction, and gaseous ammonia is extremely irritating. Clearly there was a need for a refrigerant gas that would be odorless and nontoxic, and "Boss" Kettering, representing General Motors Frigidaire Division, asked Robert McNary, Thomas Midgely, and Albert Henne to conduct a search. They knew the physical attributes needed for such a purpose, and their search of the literature brought them to a compound of carbon to which were attached two fluorine atoms and two chlorines. However, the prevailing view was that fluorine compounds were too toxic to be considered. All of the fluorine compounds known at that time were inorganic, i.e. the fluorine atoms were attached to metals, but these were of the organic type (i.e., the fluorine was bound to carbon atoms). Midgely felt they should proceed with the project, and once they had a gaseous fluorine compound in hand, determine whether it was toxic. Here is where serendipity entered the picture.

Roy Plunkett

To produce CCl_2F_2, it was necessary to react carbon tetrachloride, hydrogen fluoride gas, and SbF_3 (antimony trifluoride), but the latter was in extremely short supply. One supply house had all that was in the country, which was five one-ounce bottles. DuPont bought the five bottles and used one of them in a test run. The product appeared to be what had been expected, so the next step was to determine whether it was toxic. Accordingly, a guinea pig was placed in a bell jar and some of the Freon (as we have come to know it) was passed into the jar. The guinea pig showed no sign of distress. Someone suggested that they had better make sure by repeating the test. The second bottle of antimony trifluoride appeared to produce the same product, but when it was passed into the bell jar, it was toxic to the guinea pig. The same toxicity appeared with the product from the third bottle. How could this be? Obviously, there would be a large demand for a refrigerant that was odorless, but not if it were toxic. An examination of all of the factors revealed that the one bottle of antimony trifluoride that had given a benign product was anhydrous. All of the others contained waters of crystallization, and they produced phosgene as a by-product. It was phosgene that was killing the test guinea pigs. But, just think! If those researchers had chosen for the first test any bottle of antimony trifluoride that contained waters of crystallization, the resultant product would have killed a test guinea pig, and DuPont might have discontinued its research on fluorine compounds. We might not have ever heard of Freon or any of the other valuable compounds containing fluorine.

Because of this accidental occurrence, DuPont continued its research on fluorine, thus opening the way for Roy Plunkett to discover Teflon serendipitously. DuPont developed five refrigerant gases made of various combinations of carbon, fluorine, and chlorine. Because of patent considerations they were not allowed to market the one containing two carbons, two chlorines, and four fluorines; they decided perhaps the replacement of one of the chlorines with a hydrogen would still be a suitable refrigerant gas and avoid the patent problem. To make such a compound, it was necessary first to make tetrafluoroethylene (TFE). Therefore, Plunkett synthesized about one hundred pounds of TFE and stored it in one- to two-pound quantities in small steel cylinders that were kept at dry ice temperature. On the morning of April 6, 1938, when Plunkett's assistant, Jack Rebok, opened one of the cylinders that had been used the previous day, nothing came out. Fortunately they knew the weight of the cylinder from the day before, and when they reweighed it, found that the weight was not changed. The gas had not leaked out, but it must have been transformed in some way. When they cut open the cylinder, they found a powder that we know now to be Teflon. It had formed spontaneously by polymerization of the tetrafluoroethylene.

In telling of his wondrous experience, Plunkett was most grateful for several reasons. First, he felt lucky to be alive because it was conceivable that a sudden polymerization of TFE could have led to an explosive reaction. Second, he was grateful that Teflon had played such an important role in our society. Its first application was in making gaskets that were ideal for certain applications in the Manhattan Project during World War II. Many other uses followed (its use in frying pans is discussed in chapter 13 of this book). Plunkett ended his talk with this story:

> It has been exciting to see something I worked on become of such great benefit to mankind in general as well as to real people I know. Some years ago, at a social affair, I was introduced by a doctor friend to a man who had been suffering from a serious heart defect. "See that fellow dancing over there?" asked my friend. "He's alive today because he is wearing a Teflon aorta which I installed." Over the years I have heard of and experienced many incidents like that. It sort of makes the occasional discouragements of research a lot easier—I believe I've been more than just helpful. I've made a real contribution.

Observing a Difference

It is too early to tell whether an observation made in Japan in 1930 might lead ultimately to a cure for cancer, but the consequences of that finding have been encouraging. Bear in mind that cancer can consist of an uncontrolled rapid growth of cells; therefore, something that seems to cause such a symptom is of interest. Thus it was that an accidental discovery by a plant pathologist in Japan was so important. While walking through a rice paddy he noticed that a number of plants were twice as tall as the others, and when he looked at the roots, stems, and leaves of the taller plants he invariably found a pinkish fungus. Subsequent study determined that the cause of the rapid growth was the presence of gibberellins, which are now well-known for their role as plant growth regulators.

A rare fungus produces a compound known as Taxol, which has been a subject of interest as a potential treatment of cancer. Finding a source of Taxol became a priority, and it has been found by Andrea and Donald Stierle of Montana State University in the inner bark of a yew tree in northern Montana. Subsequently the structure of the compound has been determined, which offers the hope that it might be produced synthetically if the clinical results are sufficiently promising.

The Virtue of Not Having Any Equipment

Occasionally the very lack of equipment has been the factor that led to a beneficial change in a career. Certainly, if a capable researcher is confronted with frustration born of a shortage of equipment, he must do something about it. In such a situation many years ago Lord Rutherford said, "We have no money, therefore we must think."

One of the most prominent names in bioluminescence is Beatrice Sweeney, who did much of her work at the University of California, Santa Barbara. A postdoctoral fellow with her at Harvard was Folke Skoog, who helped her in her laboratory studies. They discussed research as they were eating cherries and throwing the pits out the window at a chimney nearby. During the 1950s she was working at the Scripps Institution with Francis Haxo on the action spectra for photosynthesis in algae. For studies of photosynthesis in red and brown algae, she needed an integrating sphere, lights, and a monochromator. But money was scarce, and her friend Marston Sargent said to her, "While you're waiting for money, Beazy, why don't you see if you can grow some of the dinoflagellates?" She acted on his suggestion and became a leader worldwide in bioluminescence. One of her discoveries was derived from having worked at both Scripps and the University of California. When she completed some parallel studies at Caltech that she had done at Scripps, the results were not in total agreement. Subsequent investigations showed that the lights in the temperature controlled rooms at Caltech were run at an overvoltage, and that produced more blue light than that used at Scripps.

When Richard W. Stow first went to Ohio State University, there was no laboratory available for his use, thus prompting him to do some thinking. The result was the development of an electrode that can be used for measuring the dissolved CO_2 content of solutions. He tested it first with Coca-Cola, and when he was satisfied with the results, extended the concept to more vital studies such as measurements of blood flow. Might this development have taken place if there had not been time set aside for thought? Maybe, but then again, maybe not.

Endel Tulving, in his PhD research at Harvard, studied eyeball movements and visual activity, the type of research he intended to continue when appointed to a post at the University of Toronto. When he arrived there, however, there was no equipment of the type needed nor were there any funds to purchase it. This prompted him to turn to the subject of memory (at that time called "verbal learning") because it required practically no special equipment. His accomplishments were so noteworthy that he was elected to the National Academy of Sciences.

Blame It on the Press

I wish I knew the whole story that I tell here but I learned of it many years ago, before I began to keep a file on serendipitous events. The essential facts are certain, but I cannot remember the actual locations that were involved.

According to my recollection, researchers in the United States and in Scandinavia performed what were intended to be identical experiments with birds, but they obtained conflicting results. After considerable correspondence on the subject, they decided that the only resolution of the impasse would be for each group to perform its experiments with the other group's birds. Accordingly, the U.S. birds were sent to Scandinavia and their birds were sent to the United States, but the resulting studies failed to clarify the dilemma. Eventually the difference in results was determined to be attributable to the type of ink in the Scandinavian newspapers that lined the birds' cages! The lesson, of course, is that it sometimes is difficult to determine the importance of every factor that might have a bearing on an experiment.

The Importance of Location

There was a rational explanation for another case in which conflicting results had been obtained in different countries. Candace Pert and Solomon Snyder, whose work at the National Institutes of Health is referred to several times in this book, had tried to identify an opiate receptor in the brain. Their plan was to use a radioactively tagged opiate and follow its adsorption to brain tissues. Dihydromorphine was their choice as an opiate, and they commissioned the New England Nuclear Company (Cambridge, Massachusetts) to tag that compound with tritium. Rather than use brain tissue as the acceptor target, they used strips of intestines because preliminary experiments indicated that the intestines responded to opiates in much the same fashion as the brain. Their experiments were inconclusive and caused them to regard their premise as being faulty. Subsequently, however, they learned that Lars Terenius (University of Uppsala, Sweden) had been successful with his experiments that were very similar. Eventually it was learned that the reason for the success of one experiment and the failure of another concerned the tendency of dihydromorphine to be degraded by light. In the brightly lit laboratory at NIH, it degraded rapidly, but in the less intense light in the Uppsala laboratory, it retained its activity.

Scientific Insight in a Streambed

Even a mundane occurrence can lead to a significant discovery, provided there is a keen sense of observation present. In this case the scientist was J. William Costerton, professor of biology at the University of Calgary (Alberta), who fell into the icy waters of Bugaboo Creek at the foot of Snowpatch Spire in British Columbia. According to his account in a *Citation Classic,* he noticed that the pebbles of the stream were extremely slippery, suggesting that a bacterial film might be present. This would be counter to the prevailing wisdom that few bacteria are to be found in such an alpine stream. An article in which Costerton was the lead author does not mention his unfortunate fall into the water—another instance of sterile scientific writing.[3] What is particularly interesting about the article, however, is the distinction made between bacteria in such a natural setting and those cultured in the laboratory. Under sterile conditions there is no formation of a glycocalyx coating; in nature, on the other hand, bacteria form this coating that promotes adhesion to submerged surfaces such as rocks. In a rushing stream the bacterial population might be on the order of one thousand cells per cubic centimeter, but a square centimeter of rock might contain a million cells. So here we have an instance in which a fall into an icy stream promoted an understanding of the differences between bacteria grown in a pure culture and those same bacteria growing in a natural condition.

Radar in the United States

Radar in this country began with an accident of time and place at the Naval Research Laboratory in 1923, the year the laboratory opened. Albert Hoyt Taylor and Leo Young were experimenting with a high-frequency communication system and had set up a transmitter on the grounds of the laboratory, which abuts the Potomac River. After placing a receiver in a car and driving it around the grounds to determine how the transmission varied with distance, they noticed that when the car was behind a building, the signal was lost. To avoid this problem they moved the experiment several miles up the river, putting the transmitter on the north end of the Naval Air Station and the receiver on the seawall at Hains Point directly across the Anacostia River, which empties there into the Potomac. It so happened that a wooden ship, the *Dorchester,* came along and the researchers noticed that it affected the signal they were transmitting. Intrigued by this, they reported the occurrence to their supervisors, who were unimpressed: "After all, you could see the ship. What good is detecting it this way?" Young and Taylor took a broader view of the potential of their discovery and bootlegged research on the phenomenon for the next five years or so, after which the Navy actually supported the research. Just sixteen years after this chance occurrence, in 1939 the Navy installed the first radar on the battleship USS *New York.* During World War II, the Navy's radar capability played a major role in the battle against the Nazi submarines that had been so severely mauling the Allied supply ships.

The Good Kind of Cavitation

During World War II another unscheduled event had a bearing on the effectiveness of radar. A British-U.S. cooperative effort concerned the resonant cavity magnetron, invented by two Britishers at Birmingham University. The twelfth prototype was taken to Washington for a demonstration. The prototype differed from the original drawings, which showed six "cavities" but X-rays of the specimens in hand showed eight. It worked well so the Americans adopted that design. What they did not know was how lucky the choice of the twelfth prototype had been, for the eleventh had only seven "cavities" and it didn't work at all.

Gabfests as Factors in Research

A theme recurring through these pages is the inestimable value of researchers chatting informally. It is exemplified by the unusual discovery of a correlation between immunoglobulin E (IgE) and hereditary occurrences of dermatitis. The two would seem to have nothing in common, but the nexus was the association of two different groups of investigators at the University Hospital in Uppsala, Sweden. Gunnar Johansson and Hans Bennich had discovered IgE in 1967, and while Johansson was working at the blood center there he found an increased incidence of IgE in patients with asthma. At the same time Lennart Juhlin, in the Department of Dermatology, was interested in patients who had hereditary allergies and also an allergic disorder of the skin or mucous membranes, called urticaria. Johansson and Juhlin often met for lunch and discussed each other's work. Because Johansson had the methodology for determining IgE, it seemed worth trying to determine whether any of Juhlin's dermatitis patients had an unusual IgE level. They found that the IgE levels were significantly greater than normal in twenty-three of twenty-eight patients with hereditary dermatitis.[4] Normal levels were found with patients having contact dermatitis, acute and chronic urticaria, and various other skin diseases.

Ancillary Reading

Another incident involving immunoglobulin research bears on the general topic of factors that are eventually recognized as important. Steven Kessler (UCLA Medical School) had an idea that did not find particular favor among his colleagues—namely, that antigenic cell proteins could be isolated by immunoaffinity techniques. Kessler had chosen a PhD topic that concerned the characterization of immunoglobulins in lymphocytes, and his method worked as well as anybody's, but the results obtained with it were still too variable. However, on his birthday he got an unexpected break. In the wake of a particularly grueling and largely inconclusive experiment, he absentmindedly began to read an article that he had previously thought to be irrelevant to his interests. The topic was the affinity of staphylococcal protein A for IgC, and his first experiment based on the insight it provided gave him the cleanest separation of lymphocyte immunoglobulins he had ever seen.[5] His contemporaries must have been impressed because the paper he wrote on the subject has been referenced more than 1,400 times.

Keeping It Simple

High technology often requires complex approaches to problem solving, but there are obvious advantages to a simple approach if that is a possible option. Robert S. Root-Bernstein suggested that given the choice between a tedious, time-consuming, instrument-intensive procedure that is likely to succeed, and a "quick and dirty" approach based on a few test tubes but which might provide good results, he would opt for the latter.[6] Another example is to be found in a letter written by Joseph Chu to my late friend John Christman. His story concerns the refractory nature of certain man-made chemicals, particularly chlorinated pesticides. Here is a part of Chu's letter:

> In the late 1960s, several graduate students tried to isolate the organism responsible for biodegrading pentachlorophenol from an activated sludge system, but failed. They tried co-metabolism using various complicated media, and they tried to prove the degradation was involving a sequence of several types of organisms. When I got the job, I tried the simplest system (i.e., the sole-carbon source is pentachlorophenol) and isolated the bacteria.

In the publication describing the study, Chu's group gave details of a continuous culture technique for degrading organic contaminants in industrial wastewater.[7] Radioactively tagged pentachlorophenol was added to the wastewater and its degradation was evident by the radioactive CO_2 given off. Progressively greater amounts of pentachlorophenol were fed to the culture, and when the "penta" disappeared quickly, isolates of the bacteria responsible for the activity were then identified.

Another interesting event related to the subject of a sole-carbon source took place in the Chemistry Division of the Naval Research Laboratory in the late 1950s. The primary event was that a telephone pole on the island of Guam was broken in an unusual manner. It had rotted at the top. Why would a creosote-treated telephone pole rot at some point other than where it made contact with the soil? Portions of the pole were sent to Dorothea Klemme at NRL for study. Her approach was to isolate whatever microorganisms might be present, and then determine how resistant they were to creosote. Using nutrient agars containing various concentrations of creosote, she observed that there was abundant growth no matter how much creosote was present. Following this, she left out all the sugar normally prescribed for the medium and found that the organism, a fungus identified as *Hormodendrum resinae,* could use creosote as the sole carbon source!

That was not the end of the story. About that time the Navy was concerned about a sludge that formed at the fuel/water interface in Navy tankers. Whether the fuel was diesel oil for a destroyer, or aviation gas for a jet plane, the potential problem was that the sludge buildup would clog filters. In the case of aviation gas, this was serious for several reasons: a jet engine could "flame out," or the fuel gauge might give a high false reading. Therefore, a pilot might think he had a full tank when, in fact, the tank might be nearly empty. The prevailing wisdom had been that the sludge formers were bacteria, but when it became known that a fungus could degrade creosote, a further search for other organisms showed fungi among the microbes forming the sludge at the oil/water interface. As growth proceeded, the whole mix could become anaerobic, which then fostered the growth of anaerobic bacteria and, unfortunately, copious production of H_2S gas, which diminished the performance of the fuel and the morale of those within smelling distance. Rex Neihof, who studied this problem for a number of years, found that one successful approach was to add a biocide to the water to prevent the microbial growth. The most satisfactory approach, however, was to institute better housekeeping practices to reduce the presence of seawater and periodically remove the buildup at the oil/water interface.

When Avocation and Vocation Meet

Some people make their own breaks, and that surely was exemplified by Richard M. Laws, a member of the Royal Society of London. He enjoys travel and used his career to further that avocation. When he graduated in zoology from Cambridge, he had offers in diverse fields such as invertebrate zoology, fish biology, and elephant seal biology. He chose the latter because he wanted to visit the Antarctic (the subject of his PhD thesis), which then started him on a career in large mammal biology. Subsequently, Cambridge University asked him to set up a research institute in Uganda, involving him in aspects of tropical ecology, after which he founded another research station in Kenya. Wanting to move to Africa was the prime motivation in this case, but during his stay there he became an expert on the hippopotamus because of the perceived overpopulation problem with those animals. At that time, Laws had been instrumental in the decision of the National Parks trustees to adopt a culling/research program on elephants and had hired a research student for the project. When the student withdrew at the last moment, Laws filled in for him. This happy accident caused Laws to shift his attention to what he considers his best work, on the elephant. Several years later in 1968, Laws was going to accept a position in Canada to work on caribou biology, but this coincided with a decision by the Canadian government to freeze all new appointments. So, he went back to his first love—Antarctica, where almost by accident he got into his next main research. When invited to give a keynote paper on the role of mammals in the Antarctica marine ecosystem, he was led to develop an interest in the broader problems of the Southern Ocean. Have interests, will travel! How many people are blessed with the ability to master so many specialties?

The Lord Shall Provide

Among the factors driving scientific advancement is money—it can provide equipment or manpower. George F. Cahill and collaborators at the Elliott P. Joslin Research Laboratory, Harvard Medical School, happened to need people. Their project required a week of demanding regimen, and they wondered where they might ever find the right type of volunteers. Their work, succinctly stated in the title of the report, required that their test subjects undergo a week of fasting while enduring ^{14}C-glucose turnover sampling, precise water intake measurements, twice-daily Douglas bag breathing, blood sampling, precise water intake and urine sampling, and minimal bed/chair activity.[8] This was a decade before there were regulations on self-experimentation, but it would seem that the regimen was suited more as punishment for criminal activity than scientific advancement. Nevertheless, they found the ideal, dependable, honest, motivated, and financially strapped volunteers at the Harvard Divinity School in students who needed the money to go home for Christmas (three hundred dollars for the week in 1966). There were one Baptist, four Congregationalists, and one Episcopalian (we are told that his glucose values were always the highest although no causal relationship was established). In short, the results led to subsequent studies showing that the brain uses keto acids during starvation, that alanine and glutamine are preferentially released from muscle, and that insulin directly inhibits amino acid release in situ in forearm muscle. Without the divinity students' need for spending money at Christmas, the study could not have been made.

Too Many Contaminants

At first glance it would seem that researchers would welcome the opportunity to have new laboratories, but one person favoring an old room would be David

Tabor at the University of Cambridge. A physicist, he was exploring mechanisms of friction and the general field of tribology.

Tabor was accustomed to using mica as a test surface because the cleavage face of mica can be molecularly smooth over areas of several square centimeters. One of his graduate students developed a technique for gluing the mica sheets onto glass without deforming them. This allowed Tabor to use the mica surfaces for tasks that could not have been undertaken with floppy mica sheets. Among his experiments was the direct measurement of the attraction between surfaces due to Van der Waals forces. In an old laboratory without centralized ventilation, the experiments worked well, but when they moved to a new laboratory, they couldn't duplicate the results gotten previously. Vaporous contaminants from floor polish or from the ventilation equipment affected the results. They had to build a special room for this work.

Incidentally, research conducted by Tabor on the effectiveness of rubber windshield blades led, thirty years later, to the development of a tire with superior skid resistance. How's that for an unpredictable payoff?

Not Enough Contaminants

One story about Alexander Fleming, who discovered penicillin, challenges a wisdom that seems totally defensible. Consider for a moment how well established is the dogma concerning sterile conditions in microbiological research. Culture media must be sterilized before use and great precautions are taken to exclude airborne contaminations. Peer acceptance of results obtained through experimentation often depends upon the assurance that no contaminants had played a role in the results being reported. All of this preoccupation with sterility makes sense, of course, because only then can the researcher say that an observed effect was due to the organism being studied. There is a counter thought, however, based on the following event.

After World War II, Fleming made a tour of the United States, during which he was greatly honored. Included in his tour was a brand new pharmaceutical plant that featured an elaborate ventilation system, thereby diminishing the contamination of any area within the facility. Fleming was dutifully impressed but was heard to say to an acquaintance upon leaving the building, "If I had been in a laboratory like this I never would have discovered penicillin." It was one of the most important medical discoveries ever made but could not have occurred under the conditions so stringently imposed in a new state-of-the-art pharmaceutical plant.

The Advantage of Being Naive

In a *Citation Classic* regarding "Immune Response and Mitosis of Human Peripheral Blood Lymphocytes in Vitro" is an interesting commentary by the author, Kurt Hirschhorn:

> I am personally convinced that the work was done and that it succeeded because, as geneticists, we naively pursued an observation in another discipline, immunology, without the full realization that our results would question established dogma. In the years since then, I have consistently encouraged students and fellows not to fear a fresh viewpoint and to use their techniques in other fields. I believe that this paper has been frequently cited (555 times) because the various preliminary findings reported attracted many proper immunologists to use a simple technique of cell culture for the study of numerous immunologic phenomena.

Methionine and Blood Clots

Despite the most elaborate preliminary literature search by someone beginning a new study, there still might be a roadblock that is both unseen and insurmountable. Such was the case of L. H. Newburgh and his associates at the University of Michigan Medical School (Ann Arbor, Michigan) when they attempted to resolve a conflict over the causes of arteriosclerosis in 1925 (note that *arteriosclerosis* denotes an abnormal thickening and hardening of the walls of the arteries, whereas *atherosclerosis* is a form of arteriosclerosis characterized by deposits of fatty substances within the arteries).

Some researchers were citing cholesterol as the controlling factor, and others considered protein intake as the likely villain. Newburgh's position was that protein was largely responsible for arteriosclerosis, and to establish the point he fed two groups of rabbits with either a thirty-six percent or twenty-seven percent protein diet. His team found that the "occurrence and extent of atherosclerosis" was roughly proportional to the number of days the rabbits spent on the diet; the difference between the thirty-six percent and twenty-seven percent protein was not significant. They were forced to conclude that protein intake was the real problem. They were sufficiently open-minded to investigate the role of cholesterol and confirmed that diets high enough in cholesterol could cause arteriosclerosis, but they discovered no significant relationship.

It so happened that the thirty-six percent and twenty-seven percent protein in Newburgh's experimental diets were much lower concentrations than the cholesterol which other investigators used to produce arteriosclerosis. Newburgh's group then took one step further in the study to determine whether some substance in animal protein might cause arteriosclerosis. Knowing that proteins consist of long chains of amino acids, they made diets based on the fifteen amino acids known at that time. The results they obtained indicated that no effect could be attributed to any of those tested, but there was a significant omission in their research.

They did not have access to amino acids containing sulfur and one of these, methionine, probably would have had a large effect. This is the contention of Edward R. Gruberg and Stephen A. Raymond who wrote *Beyond Cholesterol* in 1981. The evidence they cite is credible. Gruberg and Raymond believe that cholesterol and other fatty components can play a role in the formation of blood clots, but the more necessary components are endothelial cells released from the interior lining of the arteries. These cells form a matrix around which fatty components of the blood aggregate and form clots; in the absence of endothelial cells, the fatty substances remain suspended in the blood without ill effects. Vitamin B6 (pyridoxin) tends to inhibit the release of endothelial cells, but methionine promotes it. Many foods contain both so, for this purpose at least, it would be desirable to ingest a high ratio of vitamin B6 to methionine. Bananas have the highest ratio (15:1) and apples are second, but at a much lower ratio (5:1). The important point to note from this narrative is that Gruberg and Raymond were rigorously logical in their work and called the right shots; Newburgh was less fortunate, having been born too soon to know about the sulfur-containing amino acids, particularly methionine.

Diet and a Pain-Killer

Good planning is required in research, but there are times when the planning can go for naught without a favorable assist from the element of time. Attallah Kappas (Rockefeller University, New York) tells of a nutrition study that yielded interesting results because of its timing.[9] The study was one of a series of experiments designed to explore the influence of nutritional factors on the metabolism of pain-killers, in this case phenacetin. They had found that feeding a diet of charcoal-broiled meat to humans enhanced its metabolism, and for the present experiment human volunteers were fed a controlled diet during the study period, then given a therapeutic dose of phenacetin (900 mg). There ensued a period of time in which the phenacetin level in the blood was monitored for the purpose of determining its half-life. One of the volunteers had planned to go on the Atkins diet, which permits eating protein but drastically reduces the intake of carbohydrates. It was after this volunteer had been on the Atkins diet for two weeks that she was administered a 900 mg dose of phenacetin, and the concentration of the drug in her blood decreased extremely rapidly when compared to the others in the same study.

There was a sequel to this accidental finding (by A. H. Conney et al.) in which nine healthy volunteers were administered the phenacetin dose after each of four periods during which they were fed (1) a customary home diet, (2) a control hospital diet that contained hamburger and steak placed on aluminum foil and cooked over burning charcoal, (3) a diet containing charcoal-broiled beef that was identical to the hospital diet except that the beef was cooked directly over burning charcoal, and (4) the control hospital diet.[10] Concentrations of phenacetin in the blood plasma were determined hourly for seven hours. Phenacetin levels in the patients who had eaten the charcoal-broiled diet were consistently lower than the others. The highest phenacetin concentrations were found with the customary home diet.

These intriguing findings led Kappas and his group to conduct a major study on the metabolism of antipyrine and theophylline when dietary protein and carbohydrates were varied. Both of these drugs are rapidly absorbed when given orally. Both are substrates for liver enzymes that metabolize steroid sex hormones as well as many drugs, carcinogens, and other chemicals found in the environment. The conclusion was that the metabolic clearance rates of both antipyrine and theophylline increased significantly when the diet was changed from the normal home diet to the combination of low carbohydrate and high protein. All of these insights were gained because a human volunteer happened to be on the Atkins diet at the time of the initial study.

Nothing Is Ever Truly Garbage

An interesting account by John Hughes tells how serendipity, plus a little Scottish parsimony, provided the favorable circumstances for a discovery.[11] The chairman of Hughes's department at the University of Aberdeen, Hans Kosterlitz, was working on opiate modulation of acetylcholine release and quantitative aspects of opiate receptor interactions. He shared a common interest with Hughes on the processes affecting nerves, and it was about this time that an understanding was being developed of the specific nature of tissues serving as receptor sites for opiate drugs. Work done by Hughes and Kosterlitz showed that the vas deferens of the mouse was responsive to the various stimuli under study. When Kosterlitz retired in 1973 he invited Hughes to join him as deputy director in a drug research unit (University of Aberdeen, Scotland). Hughes had tested certain extracts at an earlier time, and though they had shown no activity, he had kept them in a frozen state, for it was his custom to disdain from discarding anything that might serve a purpose later. Six months passed before he tested the extracts again, and he found a slight activity. Subsequent investigation showed that the lack of activity initially was caused by the presence of nucleotides that had interfered with the normal role played by the peptides, the active ingredients. During the prolonged storage period, the nucleotides degraded and thus allowed the activity of the peptides to become apparent. With the help of collaborators at the Imperial College of London, Hughes was able to identify two polypeptides responsible for the activity in the extracts.

An Understanding That Developed over Time

Sometimes a long period must elapse before a new concept, even a simple one, is accepted. That is the conclusion of G. S. Fraenkel (Department of Entomology, University of Illinois, Urbana), whose studies on insect/plant relationships led to an understanding that seems reasonable but had not been formulated. Several factors had to be drawn together to facilitate the process.

In a comprehensive study made during World War II, Fraenkel became convinced that the basic food requirements of many insects were essentially identical. Furthermore, they were similar to those of higher animals. At that time Fraenkel became involved in a study of human nutritional needs that emphasized the importance of green vegetables in the diet. At an entomological congress in 1951 he presented data supporting the thesis that green leaves contained all the nutrients necessary for their insect predators in excellent quantities and proportions. There was no a priori reason why insects should not develop on any plant that they could eat. What causes them to eat one and not another, particularly if each is nutritious? Following is a part of Fraenkel's account.[12]

> By what now seems a coincidence, during the war a then-lieutenant of the Canadian Army turned up in my laboratory in England and became engaged in a PhD thesis on the role of the glucosinolates in cruciferous plants as feeding attractants for certain insects, confirming earlier results. Thus we had a situation in which all plants were potentially equally nutritious but were only very selectively eaten, suggesting a role for obviously non-nutritional plant substances in the food selection of certain insects. What could be simpler than putting these two premises together; the enormous variety in the distribution and composition of the secondary plant substances, for which no comprehensive and plausible explanation then existed, accounted for the equally stagger-

ing variety of insect/food-plant relationships, by their acting as repellents and attractants for insects (and other organisms).

There had always been speculation of the role played by secondary substances in plants, but the results found here indicated that such substances might be either insect attractants or insect repellents. In his *Citation Classic* reminiscence, Fraenkel pointed out that his article was generally ignored for more than five years judging from the few citations it received. Ultimately the citations increased and Fraenkel's interpretation was, "Perhaps it seemed implausible that such a simple explanation could be virtually new and at the same time correct."

Detecting Differences in Liquid Levels

Shortly after I joined the Naval Research Laboratory in 1956, I was involved in a study of the electrophoretic properties of fungus spores. This was a basic research study related to a problem having many roots in the Navy, and I was using a glass electrophoresis cell with a flat, shallow chamber to which were connected vertical arms in which electrodes were held by means of agar plugs. The cell was placed on a microscope stage, and the microscope was focused on that depth of liquid in which movements of fungus spores could be observed when an electric field was applied. To begin, a spore suspension would be placed in the chamber and then the electrodes would be sealed in place by carefully placing molten agar in the vertical arms of the cell. There were two problems that emerged with this procedure.

Captain Krapf, Jerry Hannan, and John Leonard

The first was the matter of sealing the electrodes in place in such a way that the spore suspension stayed perfectly still until the electric field was applied. If any minor opening was present in one of those agar seals, there would be a tendency for the whole suspension to flow one way or the other, depending upon the heights of the respective liquid levels. Any drifting of the suspension destroyed the experiment. This caused much consternation to me and my section chief, John Leonard (who also happened to be my best friend). In time, we were able to solve the problem with the agar seals, but we also came to realize that when one focuses on a particle as small as a fungus spore (about five microns), any movement of the liquid in which the spore is suspended is observable. In a typical level used by a carpenter, one looks at a bubble, but the sensitivity of that type measurement is nothing compared to the sensitivity of observing a fungus spore through a microscope. According to our calculations, the sensitivity of our method was considerably greater than the most accurate instrument used by machinists. We received a patent for the device, but I have never heard of its being used. Perhaps some reader of this book will have a need for such a device and wonder why he never thought of it in the first place.

The second problem that had to be solved was the matter of vibration. If anyone opened or closed the door to the laboratory, a vibration occurred which rendered any electrophoretic measurements useless. There had to be some way of reducing the problem, and what I did worked perfectly. I made a sandwich of three pieces of one-inch-thick plywood with two one-inch thicknesses of foam rubber. By tightening bolts connected from one piece of plywood to another, the foam rubber was compressed and when the microscope was placed on this platform, all vibrations were damped out. It worked beautifully, so when we had finished the project, this vibration-free base was stored away until it might be of use to someone else. Eventually, someone in our division mentioned that he had a problem with vibration and I jumped forward with the answer. I dug out this special base from the storeroom and watched proudly as the man tried it out. The result was a disaster! The problem was magnified by this particular vibration-free base. There seems to be a matter of frequencies involved in which I had no background, but I was grateful that it had at least worked for me.

PERSONALITIES IN THE LABORATORY

Distinguishing between Agonists and Antagonists

A research program can be dependent on personalities, and this example of serendipity shows how a difference in personalities led to an important discovery. This is a condensation of the account given in his excellent book, *Apprentice to Genius,* by Robert Kanigel (Macmillan Publ., 1986).

Candace Pert was a highly successful investigator in Solomon Snyder's laboratory at the National Institutes of Health in Bethesda, Maryland. She was miffed when Snyder took on another graduate student, Gavril Pasternak, who was to work on the same project as Pert. There was a difference in temperament between the two: Pert tended to be highly intuitive and could make good decisions based on that quality, while Pasternak was more the plodding, methodical type. A third player in this scenario was an extremely competent technician, Adele Snowman, who could go through five hundred test tubes per day, while others might normally use two hundred fifty. Snowman could work from 5:30 AM until late at night, and on Saturdays she might work on her own time.

The project they were working on concerned receptor sites in the brain, (i.e., specific sites for the attachment of specific chemicals). In the course of their work, Pasternak repeated some of Pert's work, which was annoying to her. Pasternak's excuse was that he wanted to make sure that they were both following the same procedure, but still there was friction. Pasternak was evaluating the influence of a variety of chemical reagents upon the binding of opiates, with the goal of identifying the opiate receptor. At that time, one of the compounds of interest was naloxone, which had been tagged with a radioactive tracer. When Pasternak added EDTA (a complexing reagent that can affect proteins) to his incubation mixtures, naloxone's binding to the opiate receptor doubled. When Snowman, representing Pert's team, tried EDTA in the same type of experiment, there was no effect. Since EDTA did not seem to be a terribly important reagent, Sol Snyder didn't care greatly what it did to the opiate receptor. But for Pert and Pasternak, this was another good reason for argument.

Eventually it was recognized that there were differences in the experiments they had run. Pasternak had used naloxone as the binding agent, and Snowman had used dihydromorphine; one was an *agonist* and the other an *antagonist.* As it turned out, both conflicting results were valid; EDTA decreased the binding of the agonist to opiate receptors but increased the antagonist binding. An additional problem was that when Snowman tried to repeat the result on different days, she had difficulty doing so. Sometimes EDTA didn't affect opiate receptor binding at all.

Ultimately, the problem was resolved—two different forms of EDTA were being used. One was a sodium salt, and the other potassium. The sodium salt lowered the binding of the agonist and increased the binding of the antagonist. Snowman's failures had occurred when she used the potassium salt. This unforeseen event proved to be a great blessing, because it showed that there was a way to differentiate between opiate agonists and antagonists. Had there not been a personality conflict at the outset, it is possible that this major finding would not have occurred.

REFERENCES

1. Cousins, Norman. 1989. *Head First: The Biology of Hope.* New York: E. P. Dutton. 1989.

2. Gell-Mann, Murray.1987. *Science* 132:298.

3. Costerton, J.W. et al. 1978. How bacteria stick. *Scientific American* 238: 86–95.

4. Juhlin, L. et al. 1984. *Current Contents.* July 9, pg 14, referring to Juhlin L. et al. 1969. Immunoglobulin E in dermatoses; levels in atopic dermatitis and urticaria. *Arch. Dermatol.* 100: 12–16.

5. Kessler, S. W. 1983. *Current Contents. (LS),* March 28, pg 20, referring to Kessler, S.W. 1975. Rapid isolation of antigens from cells with a staphylococcal protein A-antibody adsorbent; parameters of the interaction of antibody-antigen complexes with protein A. J. *Immunology.* 115: 1617–1624.

6. Root-Bernstein, R. S. 1994. *R & D Innovator.* May issue.

7. Chu, J. P., and E.J. Kirsch. 1972. Metabolism of pentachlorophenol by an axenic bacterial culture. *Appl. Microbiol.* 23: 1033–1035.

8. Cahill, George F. Jr. 1984. *Current Contents No. 42,* Oct. 15, referring to G. F. Cahill Jr. et al. 1966. Hormone-fuel interrelationships during fasting. (Reference cited in J. Clin. Invest. was inaccurate.)

9. Kappas, Attalah. 1987. *Current Contents.* July 6, pg 14, referring to Kappas, A. et al. 1976. Influence of dietary protein and carbohydrate on antipyrine and theophylline metabolism in man. *Clin. Pharmacol. Ther.* 20: 643–653.

10. Conney, A. H. 1987. *Current Contents,* July 6, pg 14, referring to Conney, A. H. et al. 1976. Enhanced phenacetin metabolism in human subjects fed charcoal-broiled beef. *Clin. Pharmacol. Ther.* 20: 633–642

11. Hughes, John. 1982. *Current Contents (LS)* Sept. 20, pg 20 referring to Hughes, J. et al. 1975. Identification of two related pentapeptides from the brain with potent opiate agonist activity. *Nature.* 258: (5536), 577–579.

12. Fraenkel, Gottfried. 1984. *Current Contents.* March 12, pg 1, referring to Fraenkel, G. S. 1959. The raison d'etre of secondary plant substances. *Science.* 129: 1466–1470.

CHAPTER 2

THE ADVANTAGES OF BEING LUCKY AND OBSERVANT IN CHEMISTRY LABORATORIES

"The paths of research rarely lead in straightforward fashion from starting point to desired goal. Although intention predisposes the route, chance, or occurrences along the way often enforce a change of course."

—George Wittig

All matter is chemical, therefore it is not surprising that innumerable interactions of chemistry with other disciplines can be found. It is also to be expected that serendipity would play its usual intriguing role in the outcomes of such interactions. The first tale in this chapter is so astounding that the normal reader will question its credibility, but every aspect of it has been checked.

Chromatography: How It Began

At the American Association for the Advancement of Science (AAAS) symposium on serendipity in 1986, Dr. John Christman of Loyola University in New Orleans described the events of an historic afternoon at Cambridge University. While working on a project for the British Wool Institute, A. J. P. Martin had run into a seemingly insoluble problem. The Institute had asked that he determine what amino acids were present in wool; he had been able to hydrolyze the wool (i.e. react it with water), but the resultant reaction mixture contained an assortment of amino acids that defied separation. No matter what techniques he applied, he was not able to crystallize any pure amino acids because their solubility characteristics were so similar.

Martin was in the faculty lounge at Cambridge when he was joined by another member of the faculty who asked the polite question, "How are things going?" Martin tried to explain the dilemma with which he was struggling, and in order to make his narrative intelligible, he took out his fountain pen and drew a diagram on a paper napkin. Christman explained that during the course of that conversation a corner of the napkin fell into the saucer of a coffee cup, and capillary action brought the liquid across the ink drawing. As the liquid spread, it was apparent that the pigments of the ink were being partially separated. Martin immediately recognized this separation as the germ of a great idea. He rushed back to his laboratory, where he had three pure amino acids. Taking a crystal of each, he dissolved them in a single drop of water, placed the drop on a napkin, and eluted it with water for several minutes. Because the solution was colorless, there was no visible separation of its components, but when he sprayed the napkin with ninhydrin reagent (which gives a color with any amino acid) there were three zones of color that could be seen. He had separated the amino acids by what we now know as chromatography! Subsequently he worked with the remaining known amino acids, using column chromatography and countercurrent techniques, and for that he was awarded the Nobel Prize in 1952 (with J. P. Synge).

By 1951 Martin realized that there was potentially a great field for gas chromatography, but the only detector available (the gas density balance) lacked the necessary sensitivity. Martin and A. T. James had demonstrated the separation of fatty acid methyl esters by the chromatography technique, but there was still room for improvement. Among those he asked for inspiration was James Lovelock, a multitalented person who was a medical doctor and also had been given an honorary doctor of science degree by London University. In 1956 at the National Institute for Medical Research in London (Mill Hill), Lovelock collaborated with James on the gas chromatographic analysis of fatty acids found in lipids from proteins and blood cells.

Lovelock knew nothing about chromatography until his acquaintance with Martin, but he looked back through his own experiences to see if there might be some unrecognized factor worth exploring. It so happened that in the 1940s he had been involved in a study of the comfort factors in a room, specifically the cooling effects of a slight draft. To measure such a low wind speed, he had made a novel detector that had as its principal component a small brass ball, one centimeter in diameter, coated with radium that he had retrieved from the instrument dial of an abandoned aircraft. Surrounding this ball were three circles of wire to form an open sphere. To quote Lovelock's account in *75 Years of Chromatography—a Historical Dialogue:*[1]

> The ions produced by the alpha radiation from the radium were collected on the outer rings by polarizing the space between them and the ball. Any air movement through the cage carried away ions which might otherwise have been collected.

With this instrument he could detect wind speeds as low as one foot per minute!

The drawback of this anemometer was that it was extremely sensitive to a wide range of substances, particularly smokes. Therefore it never succeeded as a detector for gas chromatography, but the principle forms the basis for smoke detectors now used worldwide.

Lovelock was aware that the greatest disturbance of gaseous ionization was the presence of substances that attached electrons, so the first ionization detector he tried was based on electron capture. It was a simple ion chamber similar to the cross-section detector used by H. Boer but operated with nitrogen as a carrier gas and with an applied potential of only a few volts (a cross-section detector required the use of several hundred volts). Lovelock's account describes in detail a succession of trials, but the narrative is shortened here to say that mixtures of chemical solvents could be separated but sometimes with erratic results. Later it was learned that carbon tetrachloride (CCl_4) vapor was slowly desorbed from silicone rubber connecting the chromatographic column to the detector. After cleaning up the system assiduously, and seeing a set of off-scale peaks developed from a 0.1–microliter volume of fatty acid methyl esters, Lovelock called Tony James, who brought a sample of allegedly pure methyl octanoate. This provided many peaks but none of them had the retention time of methyl octanoate. It was realized subsequently that the electron capture detector is not sensitive to fatty acid methyl esters, and what they were seeing were probably traces of halogenated compounds that were present as impurities.

In moving on to see if other ionization processes might be suitable for detection purposes, it became apparent that one would want to ionize the solute molecules but not the carrier gas. Nitrogen ionizes when exposed to 15 electron volts, but most organics are ionized at less than 12 eV. Lovelock remarks at the lucky break that occurred when he ordered a new cylinder of nitrogen from the storeroom and none was available; the only gas available was argon. With argon as the carrier gas, he applied a potential of seven hundred volts. With a sample consisting of fatty acid methyl esters, he was able to develop an excellent chromatogram having large peaks and a noise-free baseline. When the argon cylinder became empty, and he switched back to nitrogen as the carrier gas, there was a much less satisfactory result.

Lovelock was invited by S. R. Lipsky of Yale University to spend six months in his laboratory, and it was during that period that they developed the electron capture detector. By 1960 this was available to organic chemists throughout the world, and its ability to detect chlorine at extremely low concentrations led to the awareness that organochlorine pesticides and their degradation products were to be found all over the world. This fact ultimately inspired Rachel Carson to write *Silent Spring*, and resulted in the environmental movement that has had such enormous consequences.

To sum up, A. J. P. Martin's conversation in the faculty lounge led to paper chromatography, countercurrent chromatography, electron capture detectors, gas chromatography, smoke detectors for buildings throughout the world, and the whole environmental movement.

The prior paragraph begs disbelief, and I set about to check each facet of the narrative. Nowhere in the literature could I find a reference to the account of the paper napkin in the faculty lounge at Cambridge, so I asked John Christman where he had gotten the story that he told at the 1986 AAAS symposium. He told me that Martin had told it to him in Christman's own living room. Still, there were details that should be checked, and since many of the studies referred to here were conducted at Mill Hill, I asked permission of the personnel at Mill Hill to give a seminar there on the general subject of serendipity. Included in my presentation was the story on chromatography. About forty graduate students formed the bulk of the audience, but there were several long-term professionals in attendance, and the only correction of my account that they made was that it was probably the saucer of a teacup, not a coffee cup as in Christman's account, that led to the initial chromatographic separation of the ink pigments.

Mill Hill has been the scene of many legendary accomplishments, and I learned to my surprise that it also has a bar that is available to its staff. In the interest of scientific accuracy, I cannot make the claim for a cause and effect relationship between a bar and scientific breakthroughs, but perhaps further study is in order.

One more item requires mention here, and that is the element of timing. In another chapter in this book, instances are given in which time has been a major factor in a discovery. The same can be said of chromatography. A key event was the separation of the pigments in the ink of a fountain pen in the early 1940s. If Martin had used a ballpoint pen, which did not appear until 1945, there probably would not have been an evident separation because many ballpoint pen inks have a single pigment.

Oh, Oh, I Dropped It!

Can there be anything more embarrassing than dropping and breaking a vial containing a precious fluid? And yet, there were several scientific advances made possible by just that.

One incident took place shortly after World War II and played a role in the use of ^{32}P for animal experimentation. According to Max Kleiber, ^{32}P (a radioactive form of phosphorus) was used at Berkeley, California, in the treatment of leukemia patients.[2] One day a bottle containing the ^{32}P solution fell and was broken, and the solution had to be recovered from the floor. The first impulse was to seal the isotope in concrete and dispose of it, because it was not suitable for an injection into a human. But N. P. Garden, in charge of waste disposal, was a Scotsman who was not in the habit of throwing things away. Thinking that it might be useful for animal experimentation, he called Dr. Reiber at the University of California at Davis, who then called Kleiber. Neither had given much thought to the use of tracers at that time, but they figured there was nothing to lose, so on December 23, 1946, they injected a cow with the isotope. This led to a symposium two years later entitled "Isotopes in Animal Experimentation." There is no telling how many similar experiments have taken place since then.

At a much earlier time the discovery of the catalytic powers of PtO_2 by Roger Adams began in the same unpromising way. In the narrative of Richard Holmes,[3] one of Adams's graduate students spilled a considerable amount of a platinum solution on the floor of Noyes Laboratory. The platinum was too valuable to abandon, so Adams directed the student to scrape the floor and treat the wood scrapings in some way to get rid of the wood and leave the platinum. What ensued was fusing the scrapings with $NaNO_3$, then dissolving the cooled melt with water. One part that remained was solid black, platinum oxide. To reduce the platinum oxide to platinum metal, the oxide was treated with hydrogen while in a mixed solvent containing, by accident, some alkene (an unsaturated compound capable of reacting with hydrogen gas). To the surprise of Adams and the student, the solution soaked up hydrogen at a rapid rate and in quantities beyond that expected to react with the platinum oxide. Ever since then, one of the valuable tools available to organic chemists has been platinum oxide for its catalytic reducing power, and it all began with the breaking of a glass vial.

Three Mistakes Add Up to a Correct Answer

An example of an experiment going awry but ending on a triumphant note is supplied by James Jensen of DuPont Chemical Company (Wilmington, Delaware), who had an interest in oscillating reactions. His intent was to measure, by means of a specific electrode, the valence of cobalt during the oxidation of organic compounds. By oxidizing benzaldehyde with cobaltic ion, he could also produce a color change from pink to green. Typically an experiment would consist in heating acetic acid/cobalt solutions to about 50° C on a hot plate, then adding the benzaldehyde and observing the color change and electrode response.

Because his assistant had other duties one day, Jensen was running the experiment. Being unfamiliar with the proper setting on the heat source, he was heating it too fast and the temperature of the solution shot up to 70° C, but Jensen decided to continue with the experiment anyway. Then he reached for the reagent bottle of cobalt acetate solution and was disappointed to find it empty. He used instead a solution of cobalt bromide which was at hand. Next was to be the addition of 5 ml benzaldehyde, but there were no clean 5-ml pipettes, the type they had been using. That forced him to use a 10-ml pipette to add the benzaldehyde. With three obvious errors behind him, Jensen imagined that the experiment was doomed, but within a short time the color changed from pink to brown due to the effect of the bromide on the cobalt. A few minutes later the solution was back to pink again. In fact, the reaction oscillated all night. Subsequent experimentation showed that the oscillations do not occur much below 70° C, and they don't occur without the presence of the bromide ion. Also, an excess of benzaldehyde is needed; therefore, all three mistakes were necessary for the outcome to be so satisfactory.

Reverse Osmosis

Jensen also benefited from a serendipitous event involving hollow fibers. As part of a reverse osmosis hypothesis, his team was treating hollow fibers with various chemicals in the hope that they could preserve the internal fiber structure while they were dried. In one case he observed that the drying process reduced only slightly the water flux, but it greatly decreased the salt passage of the fiber. None of the chemicals they used really helped maintain the structure during drying, but they markedly increased the efficiency of the unit in the osmosis process. Later a U.S. patent was obtained on the process. It resulted in a major breakthrough for DuPont and made reverse osmosis commercially attractive.

The Chemistry and Astronomy of Soot

Buckminster Fuller gained renown as an architect in part because he advocated geodesic domes. He will be long remembered by his architect colleagues, of course, but his name is destined for eternity in the chemical literature. Within recent years the term "buckminsterfullerenes" has been adopted to describe a new class of compounds that had not even been imagined. Essentially they are soccer-ball structures composed only of many carbon atoms—sixty or seventy per molecule. How they came to be discovered is a story related to studies in astronomy.

The absorption and emission spectra of interstellar matter had some unexplained features, particularly a strong absorption at 217 nanometers (nm) wavelength. For many years this 217 nm absorption had been attributed to small particles of graphite; several other strong emission bands were attributed to polycyclic aromatic hydrocarbons. Several scientists were interested in explanations of these absorption bands; one was a chemist (Harold Kroto, University of Sussex, Brighton, England) and one a physicist (Donald Huffman, University of Arizona). One line of research in Huffman's laboratory had been the study of the optical properties of soot formed from graphite; pure graphite electrodes were volatilized at reduced pressure in helium and gave a five percent yield of this particular type of soot. With a laser technique at the University of Indiana, Richard Smalley bombarded graphite and produced similar effects. The soot that was produced had the two absorption bands of interest, whereas previous samples had but a single absorption band. Further study showed that this was not soot of the type supposedly present in interstellar matter, but had the unusually simple formula of C_{60}. Attempts to construct a molecular model having this simple formula resulted in a soccer-ball structure; it also could explain the presence of a C_{70} compound. Official credit for the delineation of the structure must go to Robert

Curl (Rice University), but scientists at the Exxon Corporation had arrived at the same configuration. In 1996 the Nobel Prize in chemistry was awarded to Robert Curl, Harold Kroto, and Richard Smalley for their research on carbon structures having a ring configuration, of which buckyballs were a part.

Subsequent studies have shown that atoms, including those of metals, can be fitted inside these buckyballs. Practical uses of these compounds can only be a matter of conjecture at the moment, but their development has been one of the more intriguing chapters in chemical history.

While the buckminsterfullerenes are a laboratory curiosity, it has been learned that such compounds occur naturally, and an awareness of this came about serendipitously. An unusual rock with a high carbon content was being examined with the aid of high resolution transmission electron microscopy by mineralogist Semeon J. Tsipursky at Arizona State University. Tsipursky saw some curious honeycomb patterns in the midst of graphitic and amorphous material which reminded him of images he had seen earlier of buckyballs that had been made in the laboratory. Therefore, a potential connection exists among interstellar dust, laboratory-made, and naturally occurring buckminsterfullerenes in a geologic matrix.

Going back a step further in the buckyball saga reveals that the first buckyball polymer was made at the Sandia National Laboratory in Albuquerque, New Mexico. Furthermore, it came about because chemist Douglas A. Loy happened to catch a remark made at a conference. But a different group at Sandia, that of chemist Roger A. Assink and his colleagues, made a significant finding when studying a pure buckyball crystal—but this sample had two such spectral absorption lines; later it was discovered that they were not dealing with a pure crystal, but rather one in which two oxygen molecules had sneaked into the structure. Further study showed that in a system pressurized with oxygen, the buckyballs can contain as many as six oxygen molecules in the space around individual buckyballs.

An Advantage of the "Quick and Dirty Method"

A close relative to a reagent impurity is dirty glassware. That had a salutary effect in work reported by Neil E. S. Thompson (Western Unichem Technology, The Woodlands, Texas). During offshore oil drilling, a problem that can arise is that water produced in the drilling becomes emulsified with oil, destroying its clarity. Thompson was attempting to develop corrosion inhibitors for use in oil and gas facilities. To screen experimental compounds for their corrosion inhibition, he used an electrochemical method that reports instantaneously the corrosion rate in a system consisting of oil, brine, and CO_2 or H_2S. The normal protocol would be to have a series of test solutions ready and run through the procedure with each one; his own version of this operation, however, was a little different. Rather than clean up the apparatus completely after each experiment he would do a "quick and dirty" job in which the new compound was added preliminarily to get an idea of how the compound might behave. He was working with a series of dithiocarbamates as potential water-soluble corrosion inhibitors and found that some were excellent, while others were useless. At the conclusion of one test in which he had used what he termed a "lousy corrosion inhibitor and a powerful emulsifier," the mixture of ninety percent brine and ten percent oil was completely emulsified. But when he added, without any cleanup, fifty ppm of the next compound to be tested, the emulsion completely broke down and left clear water. Two distinct phases were left in a manner that he had never seen before.

Eventually Thompson's compound was tried on a difficult emulsification problem at a major oil field in Michigan, and it worked beautifully. Thompson described his findings in the September 1994 issue of *Chemtech* and gave a detailed account of the myriad factors that could have gone wrong but, by chance, did not.

It Depends upon How the Pot Is Stirred

Nicholas Turro is well-known for his research on polymers. During his studies at Columbia University on the effects of light on polymerization, Turro discovered another factor affecting his results.

Styrene can be made into an emulsion with water by the addition of a detergent, and while in the emulsified state the photo process is enhanced by the addition of small quantities of dibenzyl ketone. With stirring and with light shining on the mixture, the polymerization proceeds. Turro had used a mechanical stirrer with satisfactory results, but when he changed to a magnetic stirrer he was surprised to find a completely different situation. The rate of polymerization was three times faster, and the molecular weight of the polymer was three times greater.

Friendship as a Chemical Catalyst

A recurring theme in scientific discoveries is the role of chance meetings. Several examples follow.

The Department of Biochemistry at Cambridge, directed by F. Gowland Hopkins, was known as an organization that fostered cooperation. In addition to Hopkins's laboratory, there were divisions headed by D. Keilin and J. Barcroft. A newcomer, Max Rudolf Lemberg, was brought into Hopkins's group and given his choice of fields to explore. Lemberg had prior experience with pigments, so he continued with that work in his new post. He isolated a green pigment from the egg shells of gulls but was not sure of the structure of the compound; he called it "oocyan," and his opinion was that it was a tripyrrole. But one day Keilin told him about a green pigment in the placenta of a dog; the genesis of the conversation was that Barcroft was working on a measurement of hemoglobin in tissue, and this green pigment interfered with the assay. Barcroft had given the pigment the name "uteroverdin," but eventually it was shown that "oocyan" and "uteroverdin" were the same compound, the contrast being that in its natural state "uteroverdin" was easier to purify. Rather than a tripyrrole, as Lemberg had thought, it was a tetrapyrrole derivative of dehydrobilirubin. It would seem unlikely that a green pigment occurring in a gull's egg could also be found in a dog's placenta, but the finding was primarily the result of close cooperation among different groups in the same organization.

Knowing the activities of one's friends also benefited C. Marcus Olson in his quest for pure silicon.[4] Until the 1940s there was no way to obtain hyperpure silicon in large amounts. Silicon of ninety-nine percent purity (therefore having impurities present at ten thousand parts per million) cost less than a dollar a pound, but for electrical use the impurities must be kept to a concentration less than one part per billion.

Olson was working at DuPont, which was concerned about its supply of titanium dioxide as a pigment for white paints. With World War II in progress, it was not likely that DuPont could continue to get its TiO_2 from India. Also the supply of sulfuric acid, which was needed for the production of TiO_2, was in great demand for the manufacture of munitions. DuPont looked to its own shores for its pigments and considered using silicon as a starting point; this option would depend on preparing silicon in a pure state. The researchers knew that the color of titanium dioxide was grossly affected by trace amounts of iron and other impurities and feared that the same might be true for silicon. Here is where Olson entered the picture, for he conceived of a scheme for preparing ultrapure silicon by reacting silicon chloride and zinc vapor in an all-quartz reactor. Glass could not be used because it would melt at the temperature required for the process.

Olson's colleagues were skeptical about his approach, because there were no literature reports of reactions of silicon chloride with zinc vapor (a recurring theme of "If it's such a good idea, somebody would have tried it already"). Olson persisted and the first trial of the process produced silicon of a greater purity than had ever been made. One of the first people to learn of this was Frederick Seitz, at the time a professor at the University of Pennsylvania, who was a consultant to DuPont. Seitz was experienced in solids and in the interaction of radiant energy with crystals. He had many contacts with physicists working in government-supported projects at his own university, at MIT, and at other institutions. He was astounded at the purity of the crystal silicon that Olson had prepared and, further, he knew of the need for pure silicon in diodes used in radar devices. As a result, within one week DuPont Pigments Research and MIT had an agreement for the preparation of pure silicon for the replacement of vacuum tubes with solid state diodes. Marcus Olson had not even known that anybody besides himself had been searching for pure silicon.

Chemistry as a Bridge in Disciplines

Accidental occurrences in one field can lead to discoveries in another. Such was the case with Eugene Roberts, distinguished scientist and director of neurobiochemistry at the Beckman Research Institute of the City of Hope (Duarte, California).[5] Roberts was interested in comparing normal tissues with malignant tissues in animals, his specific concern being the free amino acids of various normal and neoplastic tissues. Paper chromatography was used to detect the amino acids in the extracts of fresh brains of mice, rats, rabbits, guinea pigs, salamanders, turtles, alligators, chicks, and humans. In all of the brain extracts he found relatively large amounts of an unidentified compound that gave a typical amino acid color upon addition of ninhydrin reagent. Small amounts of this compound were found in urine and blood, but tissues other than the brain, whether normal or neoplastic, contained only trace amounts. Further study showed that this was a relatively simple compound (gamma amino butyric acid, or GABA). For several years the presence of GABA in the brain remained a biochemical curiosity and a physiological enigma. Roberts tried to convince neurophysiologists that they should extend their planned experiments to include the effects of GABA. No one was willing to try the compound even though Roberts offered it to them, and they had nothing to lose. It was a case of "It's impractical; I know it's not a good idea." In a review of the subject in 1956, Roberts wrote:

> Perhaps the most difficult question would be whether the presence in the gray matter of the central nervous system of uniquely high concentrations of gamma amino butyric acid and the enzyme which forms it from glutamic acid have a direct or indirect connection to conduction of the nerve impulse in this tissue.

Later in the same year (1956), however, came the first indication that GABA might have an inhibitory function in the vertebrate nervous system. Japanese investigators found that topically applied solutions of GABA inhibited electrical activity in the brain. A year later GABA was found to have an inhibitory effect on the crayfish stretch receptor system, and shortly thereafter the activity in this field increased greatly. This warranted the convocation in 1959 of the first truly interdisciplinary neuroscience conference ever held. All did not proceed smoothly because the "transmitter question" agitated many physiologists, one reason being that there was so much GABA present in the brain compared to acetylcholine, which was the only neurotransmitter known at that time. Furthermore, GABA is such a simple molecule that a vital role for it was difficult to accept (but look at the simplicity of phosphoglyceric acid and how important it is in photosynthesis).

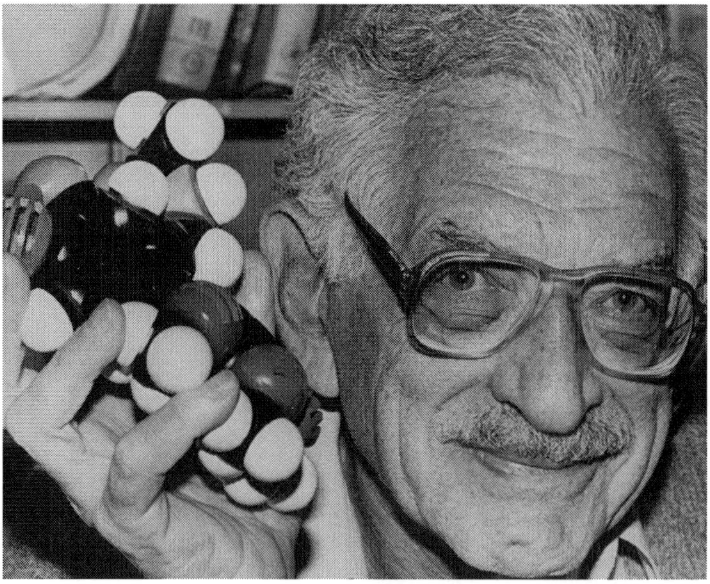

Eugene Roberts

A turning point in the recognition of GABA's importance came in the spring of 1959 when S. W. Kuffler approached Roberts and asked whether he thought there was still a chance that GABA might be a neurotransmitter. Kuffler was going to move from Johns Hopkins to Harvard and was considering putting together a team to work on the action of GABA in the crustacean nervous system. It was the work of this group, along with that of the Takeuchis in Japan, that produced the most convincing evidence for the role of GABA as an inhibitory neurotransmitter at crustacean peripheral synaptic junctions. It was a long fight for Roberts, who had not intended to thrust himself into controversy over matters unrelated to his initial investigation, but his finding had been substantiated.

Had it not been for the patience and understanding of E. V. Cowdry in the Wernse Laboratories of Cancer Research in Washington University School of Medicine (St. Louis, Missouri), Roberts would not have made such a great contribution to science. At the time he discovered GABA, he was working under Cowdry, whom he described as "a great scientist and a fine human being." For almost three years after the discovery, he was more interested in the metabolism and function of GABA in the brain than he was in cancer research, which was the mission of the laboratory. However, Cowdry encouraged him and never criticized his efforts in his research on GABA.

It is extremely discouraging for a careful researcher to meet up with invincible ignorance of the type encountered by Roberts. My good friend, Tom Rowan, had an expression that fits this problem: "To convince some people is about as hopeless as trying to teach poetry to a squirrel."

The Expanding Frontier of Nitric Oxide

For simplicity greater than GABA, think of nitric oxide (NO). In recent years this small molecule has been found to serve many bodily functions as a messenger: Gina Kolata lists its role in the regulation of blood pressure, its aid to the immune system in killing parasites that have invaded cells, its prevention of cancer cells from dividing, and its transmission of signals to brain cells.[6]

Most noteworthy is that it served as the vehicle for the 1988 Nobel Prize in chemistry that honored Robert Furchgott, professor of pharmacology at the State University of New York Health Center in Brooklyn: Ferid Murad, professor and chair of Department of Integration Biology, Pharmacology at the University of Texas Medical School, Houston; and Louis Ignarro, professor of pharmacology at the University of California, Los Angeles. These men received the Nobel Prize for their discoveries concerning "nitric oxide as a signaling molecule in the cardiovascular system." On the financial front, the role of NO became of major importance since it formed the basis for the activity of Viagra. Profits for the Pfizer Company in marketing Viagra for the second quarter of 1998 accounted for a four hundred million dollar increase over that of the year before. What is fascinating is the importance of serendipity in the various developments associated with NO.

In 1978 Furchgott was probing the relaxation of smooth muscle in isolated rabbit aorta. To shortcut an explanation of what happened, it is sufficient to say that in the course of one experiment a technician neglected to perform a necessary step and that was most fortunate.[7] This misstep led Furchgott to realize the role played in the inner blood vessel lining of endothelial cells, and he developed an hypothesis that an endothelium derived releasing factor (EDRF) relaxes smooth muscle. Credit Furchgott with a brilliant deduction, but remember that it was a goof by his technician that led to the important result.

Now let us look at the relevant work of Ferid Murad. He was at the University of Virginia when a visiting scientist in his laboratory determined that several compounds can cause the formation of cyclic GMP (or cGMP), which relaxes muscle. That led to the connection with nitroglycerin that had been known to exert its metabolic effect by the release of NO. By 1981 Murad was at Stanford, where he discovered that EDRF activates the enzyme that produces cGMP. Now the question arises: "If EDRF and nitric oxide activate the same enzyme to relax muscle, could they be related?" It was in this connection that Furchgott and Ignarro independently proposed that EDRF is the same as NO. This happened at a July 1986 meeting at the Mayo Clinic, and one of the attendees was Salvador Moncada.

Moncada pursued the matter further by adapting a machine normally used to detect NO in automobile exhausts to look for this compound in the exhalation of cells. Finding that it was produced by cells, he purchased a cylinder of NO and passed the gas over muscle cells, which then relaxed just as if they had been subject to the endothelial relaxing factor. Further study showed that blood pressure dropped when patients were dosed with NO. A body of belief holds that cells constantly release NO, thus causing the blood vessels to relax. According to Kolata's article, quoting John Hibbs of the University of Utah School of Medicine (Salt Lake City): "We had always thought that the major signals that control blood pressure were signals that caused blood vessel constriction. Now we know that the major signal dilates blood vessels."

Allied with the finding that NO dilates blood vessels was the discovery of its role in the erection of the penis. A description of the process is given by Michael Stroh,[8] who points out that within the penis are two long expandable sacs called the *corpora cavernosa.* Each has a fibrous outer wall and a spongy inner wall of smooth muscle tissue. Blood comes into the sides of the sacs from the arteries, and veins drain it out through vessels located between the inner and outer walls of the sacs. When the penis is limp, blood drains out as fast as it flows in, because the smooth muscle inside the sac is contracted. However, in an aroused male, nerves signal the smooth muscle in the penis to relax, the consequence being that the arteries feeding the sacs open up and force blood into the *corpora cavernosa.* With blood flowing into the chamber, the spongy inner walls expand and squeeze off the veins normally draining the blood and effecting an erection. This was apparently understood by the medical profession, but it was not known how nerves signal the smooth muscle in the penis to relax. According to Ignarro, the break came about in 1987 through work by Jacob Rajfer, a urologist at the UCLA School of Medicine. Rajfer had been studying impotence for years, and he had a hunch that the condition arises when the smooth muscle in the *corpora cavernosa* fails to relax. One day he was returning to his office, but he made a wrong turn and ended up at the door of an unfamiliar laboratory bearing the sign "Pharmacology—Smooth Muscle Lab." At the time, no one knew much about smooth muscle in the penis, but under a microscope it looked like any other muscle. What caused smooth muscle anywhere in the body to relax? Rajfer walked into the laboratory, which was Ignarro's, and found that it was only a few weeks earlier that Ignarro had found the answer, at least for arterial smooth muscle.

Ignarro and his coworkers had found that neurotransmitters in the blood bind to receptors on the endothelial cells lining the insides of the arteries, and then these cells begin to manufacture NO. Following this, the NO diffuses into the adjacent layer of smooth muscle cells and this diffusion prompts a chemical reaction, causing them to relax. Hearing this explanation intrigued Rajfer because the smooth muscle in the *corpora cavernosa* has an endothelial lining. Subsequent cooperation between Rajfer and Ignarro, using rabbits as test organisms, confirmed the role of NO in the whole erection process. What a cascade of findings was necessary to unravel this mystery! A study of car exhausts and of nitroglycerin, Rajfer making a wrong turn and running into the sign on Ignarro's laboratory—all these factors were necessary, and they ended up with an award of the Nobel Prize.

Finally, a word of explanation is in order regarding the connection between NO and Viagra. Viagra (sildenafil) was discovered in 1989 during basic research studies on the enzyme phosphodiesterase (PDE5) that breaks down cGMP. It was originally tested against angina but participants in a 1992 trial reported little effect on heart function; instead, their sex lives improved. According to the account by P. A. Palevitz and R. Lewis, penile nerves, when excited, release NO or stimulate endothelial cells to do so.[9] Then NO receptors on smooth muscle cells in nearby arteries activate the enzyme guanylate cyclase, which catalyzes the production of cGMP. It is cGMP that leads to muscle relaxation and that dilates the vessels. Under normal circumstances, cGMP does not last long because it is broken down by PDE5. Viagra inhibits PDE5 and therefore serves to amplify the dilation.

One cannot help but marvel at the economic consequences of discoveries related to serendipity. (1) Penicillin: how many prescriptions containing this antibiotic have been sold? (2) Post-It notes: It is said that 3M does not disclose how many are sold every year. Initially, the company saw no future in them because they didn't adhere strongly to surfaces, but when a few secretaries learned about them, the demand was enormous. (3) Teflon: how many of these household necessities have been sold since 1962? (4) And recently, the sales of Viagra and related products exceed hundreds of millions of dollars. For obvious reasons, the propriety of featuring TV ads on their behalf is understandably being severely questioned. A succession of factors contributed to the discovery of Viagra, but the role of a molecule composed of only two atoms (NO) makes it most unique.

Not All Sulfurs Are the Same

In a 1992 *Citation Classic,* Carl Nordling told of the circumstances involved in a significant discovery that resulted from the substitution of a wrong (though pure) compound in an experiment.[10] It was in the early stages of electron spectroscopy, and his team was exploring the electronic structures of atoms and molecules. The method, developed in the early 1960s by Kai Siegbahn, involved expelling electrons from their normal state by X-rays and measuring the forces binding them to the parent compound. Nordling's group was exploring the powers of electron spectroscopy for the analyses of various elements and had planned to use Na_2SO_4 as a source of Na, S, and O. This was performed in the physics department laboratory at night, and it happened that no Na_2SO_4 was available. But $Na_2S_2O_3$ was in their cabinet, and they decided to use it since it contained all of the elements they were interested in. To their surprise the sulfur component showed up as two separate peaks, and it was in this way that they came to appreciate the potential of electron microscopy for a more refined chemical analysis than they had imagined. Their article in a book on electron microscopy for chemical analysis (ESCA) has been referenced more than one thousand six hundred times, so its importance is well-established.

A Payoff from Military Research

Sometimes all that is needed for a successful study is a trivial bit of information. Henry Jeffay and Julian Alvarez were using ^{14}C to follow the breakdown of protein in the body in a project at the College of Medicine, University of Illinois,

Chicago. That sounds straightforward enough, but to determine the radioactivity of a sample, they needed a method that would provide a linear correlation of counting with radioactivity, and it had to be independent of the volume of liquid used. For more than a year they tried using a polar organic solvent but finally gave up. The breakthrough came one Sunday morning when Jeffay was reading a story about life in a nuclear submarine, and it was mentioned that the CO_2 respired by the crew was quantitatively absorbed by ethanolamine. This rang a bell for Jeffay, and the next morning he and Alvarez got hold of some ethanolamine, tried it as a trap for CO_2 they had obtained through oxidation of a ^{14}C sample, and the whole system worked. They were able to finish the project in a few days. Subsequently some changes were introduced, but they published an article in *Analytical Chemistry* that has been cited more than f our hundred and fifty times.

Once again, the Marvel of Chromatography

My good friend, John Christman, who died in October 1995, should properly be referred to as Mr. Serendipity for his encyclopedic knowledge of the subject. As a member of the Tour Speakers Bureau of the American Chemical Society, he presented 171 talks to local sections, and anyone who heard him speak would never forget his great sense of humor and the breadth of his knowledge. "Chris" was fond of telling a story of his teaching career while at Louisiana State University (Baton Rouge) involving two graduate students, Tommy Jackson and Robert Trubey. As is so often the case, the principal and most important finding had nothing to do with the original intent, which was to study the production of lactic acid by a *Lactobacillus* bacterium. As background, several groups had noticed that glycerol trioleate would support the growth of *Lactobacillus* without biotin being present in the culture medium. What would be the result of either high or low concentrations of biotin on the acid production of *Lactobacillus?* Jackson found that very little acid resulted under either circumstance. Trubey, with an analytical chemistry background, was given the task of devising a paper chromatography procedure for separating fatty acids. He found a promising answer based on the combination of a filter paper lightly coated with latex and low-boiling petroleum ether as a solvent. With this procedure he seemed to get a separation of fatty acids but their "front ratios" were not very different (i.e., the distance each acid migrated from the top of the paper). An additional problem was that when he pulled the strip from the chamber, the solvent evaporated so quickly that he couldn't tell where the "front" was. Chris suggested to him that he use a colored compound that had a "front" ratio of one and place it parallel to the sample so that he would have a marker at the actual "front" of the solvent. So what would be a proper colored compound for that purpose? Chris suggested that Trubey see Henry Werner, professor of cytology, who had an almost endless variety of dyes.

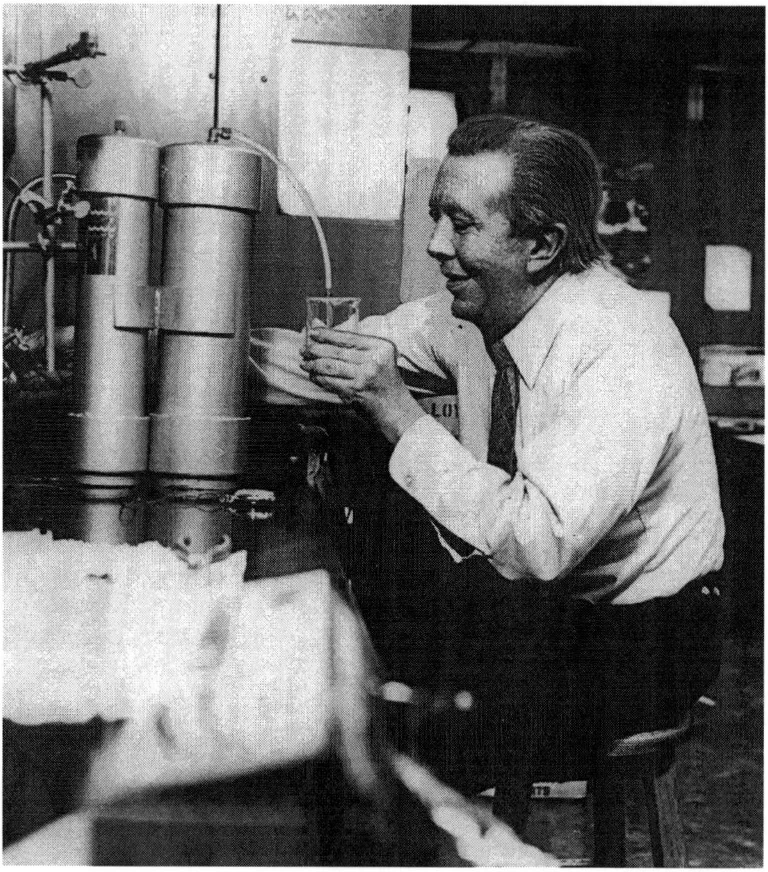

John Christman

Trubey tried to explain to Werner what his problem was, but Werner had no knowledge of chromatography at that time. He did understand, however, that the student was trying to separate fats so he gave him samples of Sudan III and Sudan IV, which are used as fat stains. Trubey then ran a chromatogram of Sudan III and expected to find a nice red marker at the top of the chromatogram. What he got was far different—about fifteen various colored bands ranging from a very pale yellow to a very dark purple. As he showed the mess to Christman, he made a motion as if to throw it in a wastebasket, but Chris's reaction was to say, "Hold on. Let's not get hasty. Maybe we can use this." Chris then called Werner and asked if he ever had any trouble in making a fat stain, to which Werner responded that everyone in that field had trouble from time to time. Chris suggested that the problem might be that there was a tremendous number of impurities in the dyestuff, but Werner's response was that the dye was marked 97.5 percent pure. When asked whether he would like a sample of very pure Sudan III, Werner just about jumped through the phone saying yes. Chris and Trubey went to Werner's office and chromatographed some of the Sudan III, using a diatomaceous earth column and petroleum ether. The major band was bright red and it separated easily from the contaminants, producing ultimately some very nice crystalline Sudan III. It was the first time this compound had been crystallized. But the pure dye didn't stain fat! In fact, it wasn't even soluble in forty percent alcohol/water, which was the solvent used for the fat stain. Obviously, whatever was doing the staining was an impurity in the dye rather than the dye itself. Despite the subsequent clear knowledge that neither Sudan III nor Sudan IV stain fat, the commercial products are still best-sellers as fat stains. Chris reported that the major fat-staining component is benzene-azo-beta-naphthol, known as Sudan R, but there are other fat-staining components in both Sudan III and Sudan IV, and the cytologists prefer the results they get from the mixture rather than from the pure compounds. One source of satisfaction was that Chris received more than six hundred requests for reprints of the paper describing the work.

While on the subject of chromatographic techniques, mention should be made of the background for a useful means of identifying peptides and other compounds containing the -CONH- group. It consists of a chlorination step which replaces the hydrogen on the nitrogen with chlorine, after which the chromatogram is sprayed with KI solution, followed by starch to visualize the liberated iodine. H. Norman Rydon described the background of the work in which P. W. G. Smith was coauthor in a *Citation Classic.*[11] Smith was working toward his PhD at Imperial College (London) on a project involving the polymerization of glycine peptide esters, and a method was needed urgently for detecting the polymerization products on paper chromatograms. It was at this point that Rydon recalled that during World War II he had investigated the N-chlorination of nylon, using gaseous chlorine, in the hope that the nylon could be made impermeable to mustard gas. The chlorination had worked well with nylon, and that was what prompted Rydon to try chlorinating peptides that were being made by Smith. Subsequently, modifications of the method made it more sensitive, but it is interesting to note, once again, the role that World War II played in a totally unexpected way.

Stabilizing Proteins in an Unexpected Manner

A story related by John H. Crowe in response to my inquiry concerned an unusual function of a trace metal component. John Carpenter, a postdoctoral fellow in Crowe's laboratory at the University of California at Davis, was conducting research on the stability of dry proteins. Before joining Crowe's group, Carpenter had found that the sugar trehalose was remarkably effective in stabilizing dry proteins, far more so than any other sugar that he had tested. When he went to Crowe's laboratory, however, he was unable to duplicate those results. While pondering what might be going wrong, he recalled from earlier communications with Sigma Chemical Company (his earlier supplier), that zinc was a trace contaminant used in the purification of the product. What he found at Crowe's laboratory was a high-purity trehalose that did not perform the same function. When he added zinc to it, the question was solved.

Included in Crowe's letter was the news that he started his research career with interests in the biochemistry of organisms that can survive complete dehydration. All such organisms produce trehalose at high concentrations, and the first preservation experiments they conducted were with Ca-transporting vesicles that were isolated from lobster muscle. Those experiments were successful, and they learned subsequently that they were lucky in the choice of lobster, because lobsters use trehalose as a natural blood sugar. Their membranes have mechanisms for the crossing of trehalose. Had they chosen another membrane as the model (Ca-transporting vesicles from rabbits, for example) the experiments would have failed and the story would have stopped there.

Incidentally, Crowe's interest in the subject of this book is evident in his outlook on life. He and his wife have a sailboat bearing the name *Serendipity.*

Problems with Reagent Purity

Crowe's experience brings to mind other examples of laboratory studies being affected by impurities. An account in *Chemical and Engineering News* concerned the unexpected presence of benzene in Perrier sparkling water. A North Carolina environmental laboratory (Mecklenburg County Environmental Protection Department) had used Perrier sparkling water as a diluent to prepare water samples for analytical tests because it was reliable and convenient. Prior to that time, according to laboratory director James T. Ward:

> We used to take deionized water, boil it, purge it with nitrogen, seal it in a container, then cool it down. But this was fairly time-consuming for the amount we were using. So about five years ago we tried the bottled water. There were no organics in it, and it worked pretty well. It was a lot easier and less expensive to use.

But on one occasion, a gas chromatograph/mass spectrometer was being used to analyze water samples for volatile organic priority pollutants such as benzene. When the operator ran a blank in the morning to check out the procedure, he got a peak on the screen. It took several days of experimentation, checking method- or apparatus-related possibilities, before they realized that the Perrier water contained benzene. Further study of fresh bottles of the product showed benzene to be present at concentrations of thirteen to fifteen parts per billion, which was three times the EPA limit of benzene permitted in drinking water.

The Food and Drug Administration then ran further tests and found similar benzene levels in Perrier water from North Carolina and Georgia. It was concluded that the levels being reported would not pose an immediate danger to health, but there would be an increased risk of cancer for lifetime exposure. Source Perrier S. A. decided to recall the product from world markets, and its theory was that the problem was caused by filters used in the carbonation process.

I had a personal experience with filters that is worth a mention. It was in the 1970s that my section at the Naval Research Laboratory had an interest in sediments that had been picked up on various research cruises. We had devised a bioassay based on the growth of *Phaeodactylum tricornutm,* and for the sediment study our procedure was to mix a sediment with artificial seawater, filter the resultant suspension, inoculate it with the test organism, and measure the growth after two days. Generally there was little, if any, difference between the growth of the controls and the test cultures made with the sediment extracts. But then there came a succession of experiments in which the sediments being tested were highly toxic to the test organism. The results were suspicious, so there ensued an extended study of all of the factors that might have influenced the results. Fortunately my assistant, Connie Patouillet, who conducted every assay with the most extreme care, checked her laboratory notebook and found that the problem with apparently toxic sediments coincided with the use of a new box of millipore filters. With this as a lead, it became apparent quickly that the filters were the cause of the problem. A phone call to the manufacturer brought a rather defensive reply, as might be imagined, but when I made the point that I really was not complaining, nor was I asking for any money back for the purchase, the tone of the conversation changed. I learned that the practice was to add a detergent to the filters to increase their wettability, sometimes even two percent being added. Microorganisms can be extremely sensitive to detergents, so the mystery was solved.

The bioassay based on the growth of *Phaeodactylum tricornutum* involved another interesting turn of events. Initially for convenience we used artificial seawater that was prepared by mixing sea salts with distilled water. Problems developed, however, and we were certain that the sea salts varied in composition from one lot to the next. We switched to seawater prepared with chemically pure reagents and distilled water, made up in fifty-five-gallon plastic drums. Included in each assay was a standard toxicant in the form of $HgCl_2$ (0.03 ppm Hg). Over the course of a year the results were very consistent, with repeatable growth rates of the organism and also with growth inhibition caused by the mercury.[12]

When the initial batches of this seawater made with CP reagents ran out, we made up new batches. Because rather large amounts of the inorganic salts were needed, we necessarily used new bottles of the reagents. Results over the next year showed differences in the growth rates of the control cultures and also in the inhibition caused by the standard concentration of Hg as a reference standard. Further study showed the surprising concentration of mercury that was found in the control cells. They contained fifty-two ppm mercury. The most likely source of that was the chemically pure reagents used in the artificial seawater, even though the labels of each made no mention of mercury being present. Of course, algal cells tend to concentrate mercury enormously (the amount depending upon the concentration of mercury present) so the mercury concentration in the culture medium might well be below the detectable limit. I can only wonder at how many studies have been made involving microorganisms where the concentration of some inorganic component might be relatively large without the experimenter having any idea of its magnitude.

REFERENCES

1. Ettre, L. S. and Zlotkis, A. ed. 1979. *75 Years of Chromatography*. Elsevier Publ. Co.

2. Kleiber, Max. 1967. An old professor of animal husbandry ruminates. *Ann. Rev. of Physiology*. Vol. 29, 1–20.

3. Holmes, Richard. 1997. The origin of serendipity. *Chemical Heritage*. 14: 43.

4. Olson, C. Marcus. Spring/Summer, 1988. *Invention and Technology.*

5. Olsen, R. W. and Venter, J. C. eds. *Benzodiazepine/GABA receptors and chloride channels*. New York: Alan R. Liss Inc., 1986. Chapter 1. GABA: The road to neurotransmitter status. Pgs 1–39.

6. Kolata, Gina. July 2, 1991. *New York Times.*

7. Lewis, Ricki and Palevitz, B. A. 1998. *The Scientist.* 12: 1–4.

8. Stroh, Michael. 1992. The root of impotence. Does nitric oxide hold the key? *Science News.* 142: 10–11.

9. Ref. 7 *loc cit.*

10. Nordling, C. 1992. Making electrons talk. *Current Contents,* No. 20, July 27, referring to Siegbahn, K. et al. 1967. Atomic, molecular, and solid-state structure studied by means of electron spectroscopy. *Nova Acta Regiae Soc. Ups.* 20: 1–282, 1967.

11. Rydon, H. Norman. 1987. *Current Contents (LS), Vol.* 30, pg 15, referring to Rydon, H.N. and Smith, P. W. G. 1952. A new method for the detection of peptides and similar compounds on paper chromatograms. *Nature.* 169: 922–923.

12. Hannan, P. J., and Patouillet, C. 1972. Effect of mercury on algal growth rates. *Biotech and Bioengrg.* 14: 93–101.

Chapter 3

Soaring, Bubbling, Lightning—Where Air Meets Sea

"The sea never changes, and its works, for all the talk of men, are wrapped in mystery."

—Joseph Conrad

Anyone involved in oceanographic research can attest to the difficulty of understanding and predicting the physical processes in the sea, but much has been learned about air/sea interactions.

We begin with Albert Woodcock, whose unusual beginning as a scientist is detailed in chapter 12, "Chance and Finding One's Niche." He had little training when he signed on as a deckhand for the *Atlantis,* which was being built for Woods Hole Oceanographic Institution, but subsequent events showed him to be an alert observer. Early in his career he studied the soaring of herring gulls, using two simple pieces of equipment (i.e., a thermometer and an air speed indicator). With these two simple devices, he made sense out of what had been regarded as random flights of seabirds. Woodcock found that there are two types of organized convection in the subcloud layer at sea. When the air is warmer than the water, the gulls are content to sit on the sea surface, but when the air is colder, the gulls take off and soar without difficulty. Furthermore, he was able to identify two different types of soaring. With wind speeds less than about fifteen mph the birds fly circular patterns and slowly move upward, which Woodcock interpreted as being caused by organized updrafts. With wind speeds over fifteen mph the soaring routine changes to what he described "as nearly two-dimensional as it is possible for flight to be." The birds soar directly into the wind in a thin vertical sheet, rising and finally disappearing from sight. If the wind increases to about thirty mph the gulls do not soar, leading Woodcock to conclude that organized convection has broken down. This is one piece of the story which we will leave for the moment, but the reader is left with the reminder that Al Woodcock, with no scientific training, had made valid observations about winds at sea simply by watching birds.

Now, fast-forward to World War II when Woodcock worked briefly with a team charged with the task of hiding the wakes of the Navy's amphibious landing crafts. Even at night these ships could be detected from the air by their luminous wakes. Woodcock spent many hours evaluating the performance of various film-forming compounds as foam breakers but to no avail. His next approach was to prevent detection of the wakes by creating smokes. Because Woodcock had learned so much about convection currents in his bird-watching days, he was asked to participate in the smokescreen project. A group of ships and planes in the Gulf of Panama had encountered difficulties with fog/oil generators. Sometimes the smoke they made worked beautifully, hugging the sea and hiding all the ships, but often the smoke went straight up. Woodcock's analysis was that the generators were not at fault but rather the instability of the atmosphere. As the birds had taught him by their flight patterns, Woodcock knew that when the sea is warmer than the air, the smoke will move upward and not spread out.

Why had fog generators been developed in the first place? The story began before Pearl Harbor when the Chemical Warfare Service asked Irving Langmuir and his assistant, Vincent Schaefer, to study how gas masks work. Charcoal is used as an adsorbent for poison gases, but what would happen if the enemy used smokes rather than gases? The concern was that they might remain suspended in the air as dust particles and therefore not be adsorbed by charcoal; or that their removal would require something akin to a filter paper. To test the effectiveness of the masks, Langmuir and Schaefer had to make smokes consistently with well-defined particle sizes. After a year of experimentation, the distinguished team of an older theoretician (Langmuir) and a younger experimentalist (Schaefer) developed a procedure for making a nonvolatile smoke consisting of very small particles and had learned how to measure its optical properties. Once again, remember that the smoke being considered to hide ship movements had as its origin a study of gas masks, and that Woodcock's understanding of smoke movements had originated from his observations of soaring birds.

We pick up the narrative with Woodcock's involvement in the summer of 1947 with the "red tide" problem off the west coast of Florida. Periodically an enormous population of algae develops, which turns the sea red and causes ecological havoc. Many fish die and wash up on the shore, and seaside residents often experience a burning sensation in the nose, throat, and lungs. Woodcock was asked to work on the problem and with two companions spent much time in a small boat studying the pesky red tide. According to his account, "For long periods of time we worked over tubs of this water, over nets slimy with concentrated plankton, and decks awash with spilled red water. We noted no respiratory trouble of any kind." Soon after, a thunderstorm passed over and produced high winds and rain. When the sea was streaked with foam, Woodcock and his co-workers experienced coughing and a burning sensation in the nose and throat. The conclusion drawn was that tiny droplets were tossed into the air from the foam, and were then breathed in through the nose or mouth. To confirm the theory, Woodcock scooped up a bucketful of the red water, passed bubbles of air through it, and inhaled the air above the bucket. Sure enough, he experienced the nasal and throat discomfort characteristic of that suffered by residents of the area. As a curious sidelight of this study, the senior scientist who wrote the final report for the U.S. Fish and Wildlife Service made no mention of this phenomenon.

Woodcock went back to his laboratory and shortly thereafter met Duncan Blanchard who was finishing a master's degree in physics at Pennsylvania State University (University Park). Blanchard's improbable entry into physics is described in chapter 12. Woodcock invited him to come to Woods Hole in March 1951, where his help was sought in the measurements of raindrop sizes. The two discussed whether large particles of sea salt, which Woodcock had found high in the air, acted as nuclei around which raindrops could form. They studied how salt particles were produced at the surface of the sea, thus reopening the interest in bursting bubbles, which had attracted Woodcock's attention with the red tide. One of their laboratory exercises was to watch a single small bubble rise to the surface of seawater; it simply disappeared, but in the air above the spot where the bubble had broken, four or five drops appeared. This observation prompted Woodcock and Blanchard to determine in detail what was happening. By high-speed photography they saw that when the top of a bubble broke, the air cavity within the bubble began to collapse. Then from the bottom of the cavity a pencil-like water jet moved upward rapidly and broke into four or five droplets that continued to coast upward. The process took only a few thousandths of a second, and the speed of the drops being formed by this upward jet motion was sometimes more than one hundred ten miles per hour!

Al Woodcock, Duncan Blanchard, and Ramon Cipriano

Bill Garrett and Bill Barger

One day they noticed that the heights to which the drops rose varied systematically even though the bubble size did not change. They wondered whether thin films of organic matter on the water surface could influence the pencil-like jets from the bottom of the bursting bubbles; further study showed that as the bubbles broke, they skimmed off portions of the surface. Later they were to learn from Bill Barger and Bill Garrett at the Naval Research Laboratory that droplets rising from the sea contain concentrations as high as three thousand five hundred parts of organic matter per million parts of seawater. By contrast, bulk seawater might have only ten parts per million. All these considerations cast a new light on

Woodcock's previous observations of red tides. If the algae produced noxious compounds that migrated to the seawater surface, they could easily be released into the atmosphere by the breaking of bubbles. This explained why Woodcock could be working in the midst of the algae-laden waters with no ill effects, but experience difficulties in breathing when winds whipped up the surface to produce a foam. While the origin of the red tide may not be completely understood, there is no doubt about the mechanism by which it bothers residents of the seashore.

Woodcock's preoccupation with the atmosphere caused him to wonder at the marvels of clouds formed by the contact of boiling volcanic lava with the sea. At first glance it seems to be a straightforward process, with water coming into contact with a very hot river of lava, being vaporized, then condensing as the vapor reached the cooler areas of the atmosphere. But is it just water vapor that is involved? Woodcock was fortunate to have talked with an observant native who had seen steam clouds produced when lava from the Hawaiian volcano Mauna Loa flowed into the sea. This person remarked that the paint on automobiles in the area was spotted in a peculiar fashion. If the clouds were essentially distilled water, that could not be the case. But what if there were sea salt particles suspended in the air? Salt is a great destroyer of paint films and its deposition on a car would have obvious consequences.

In 1960 Woodcock took advantage of an eruption of Kilauea to study the problem. He rented a small plane, flew into the clouds above the volcano, collected particles on small glass slides, and examined them under a microscope. He found sea salt particles in numbers he could never have imagined. In earlier times Woodcock had actually collected sea salt particles from a tower at sea during a hurricane, but the numbers there were nothing compared to what he found in the steam clouds formed by Kilauea. Obviously, when hot lava hit the sea it did not just evaporate water—it resulted in an abundance of sea salt in the atmosphere.

Woodcock's interest in the phenomenon prompted him to try to duplicate it in the laboratory. With the aid of an associate, he heated a chunk of lava with an oxyacetylene torch until it was a molten puddle, then squirted seawater onto it. The resultant cloud was allowed to rise through an open-ended cardboard cylinder, then was trapped by placing covers on both ends. Clean glass slides on the inside of the bottom cover became coated with tiny particles. The molten lava had produced sea salt particles at the rate of one hundred million per square centimeter per second! In his book, *From Raindrops to Volcanoes,* Duncan Blanchard made the point that this was the first time attention had been focused on sea salt particles that were produced when hot lava hits the sea.

Blanchard watched Woodcock and his associate, Spencer, perform their volcano act in the laboratory many times, and it never occurred to him that there might be something extraordinary about it other than its function in making sea salt particles.

But here is where serendipity occurred. Blanchard has a hobby of reading scientific writings of the last century, and he happened to pull off the shelf in the library the *British Philosophical Magazine* for 1841. While thumbing through it, he came upon an article by a French physicist, Athanase Peltier, concerning the electricity of steam. At that time certain mill workers in England were afraid to approach the steam engines because when in operation they were highly charged and anyone who touched them got a shock. Peltier's explanation for the phenomenon was that there was a certain amount of salt in the water in the boiler, and perhaps the splashing of the salty water against the hot walls makes an electrically charged cloud. When the charged cloud leaves the boiler, its equal but opposite charge stays behind in the boiler. Peltier sought to prove this by connecting an electrometer to a hot platinum dish and splashing a drop of diluted seawater on it. He did get a wiggle out of the electrometer needle. As soon as Blanchard read this he thought of Woodcock's laboratory study of molten lava being hit by seawater and it raised the question of electricity being formed as the result of that operation. Peltier referred to a paper by Alessandro Volta in 1782 who was interested in the idea that electricity could be generated by the evaporation of water. Many times he connected pans of water to an electrometer with no luck. One day Volta was in London to give a lecture to the Royal Society and hoped to impress the members with his evaporation experiment. The thought suddenly came to him that maybe the problem was that he wasn't evaporating the water fast enough. So he took a metal dish full of hot coals, connected it to his electrometer, and threw some water into it. To his delight the electrometer gave a strong negative reading which indicated that the cloud rising above the dish must have been positively charged.

Blanchard went back to the laboratory where he quickly showed that positive electricity was carried by the cloud when seawater was tossed onto molten lava. The cloud produced by a single drop of seawater carried about a billion elementary charges. An opportunity to prove that this could happen in a natural setting came about when the island of Surtsey was formed near Iceland by a volcanic eruption. Lightning was reported as being associated with the eruptions, lending credence to the theory that clouds formed by the action of molten lava on seawater should be electrically charged. A team of scientists, including Blanchard, visited the scene on board a fishing boat. They approached the island on the southwest side and, since the wind was from the southeast, the eruption cloud would not blow directly over them. Blanchard's book contains a graphic description of their observations:

> Every few seconds dark clouds rose from the vent on the other side of the island. The wind spread them out to form a huge dark vertical wall, which ran from the island to as far as we could see on our right, and from the sea upward to nearly the zenith. On the left we saw clear air, seabirds, and sunlight; on the right, a curtain of ash and the near darkness of night. Here and there vertical pillars of cloud broke the curtain; from far above, long fingers of ash (and maybe water) streamed downward to the sea.
>
> By now we had expected to hear the roar of the volcano, but an eerie silence prevailed. Surtsey seemed to be performing in pantomime. Then every so often a sudden sharp crack, like a rifle shot, shattered the silence. We were mystified, and for a while thought the sound might be caused by rocks falling on the island. We looked more closely but could see no large rocks crashing into the island.
>
> The boat (*Haraldur)* was now closer to the curtain of darkness, and every minute or so a flash of lightning darted out from the lower part of the clouds. A few seconds later the mysterious crack was heard, but now it was no longer mysterious. It was clear that it was thunder. None of us had ever heard thunder like this before; we were familiar with the type that booms and reverberates for many seconds, like the sound of a hundred cannon all going off at slightly different times. Surtsey's thunder did not sound like this, probably because the lightning strokes were only a thousand feet or so in length. Thus the sound from every point in the lightning channel arrived at the *Haraldur* at about the same time, producing the sharp crack. A thunderstorm lightning stroke, on the other hand, may be ten thousand feet or more in length, and the time interval between the arrival of the first sound waves and those from the far end of the lightning channel may be several seconds.

An electrometer on board the *Haraldur* measured the change of potential in the atmosphere. At the start of an eruption the potential gradient was around +100 volts per meter, a value normally found at sea. Then, as the clouds rose upward, the gradient rose rapidly to several thousand volts per meter until lightning flashed, when it would again drop to the +100 level. Blanchard's hobby of reading old scientific writings had paid off with a thorough understanding of one of nature's most spectacular tricks.

Another study of cloud phenomena had taken place earlier when Langmuir and Schaefer worked on various problems connected with World War II efforts. Among them was the problem of precipitation static. At times pilots flying into the Aleutian Islands reported that their planes would ice up rapidly, and at other times there would be only snow in the clouds. The complication provided by the snow was that it provoked static in the planes' communication systems and caused them to lose contact with their home base. Langmuir and Schaefer were enthusiastic mountain climbers and chose to study the snow problem on a mountaintop rather than in aircraft. Their approach was to expose various strips of metal to the snow-laden clouds atop Mount Washington (ambient temperatures between -10° and -30° C) and measure the charges acquired, but the surfaces exposed became covered rapidly with ice. There were few snowflakes in the clouds. How was it that the clouds were composed mostly of supercooled water droplets? There had to be some mechanism by which these droplets were transformed into ice crystals and then into large snowflakes, which would then fall from the cloud.

Schaefer, the brilliant experimentalist, went back to his laboratory and designed a rather simple experiment to study the phenomenon. He lined the inside of an ordinary home freezer with black velvet to provide a dark background. As a source of water vapor he simply used his own breath, and then directed the narrow beam from a microscope lamp through the mist that his breath had formed. Even a single ice crystal should sparkle under these conditions. Schaefer kept the cold box considerably below 0° C but never observed ice crystals. What was formed appeared to be very much like the clouds he had seen on Mount Washington. Though he tried to seed the clouds with all sorts of solid materials (e.g., sand, carbon, talcum powder) nothing worked.

Serendipity helped to solve the problem. On a hot and humid day in July 1946 Schaefer noticed that the cold box was less than frigid. One quick answer was to place a block of dry ice (temperature -78° C) in the box; when he did so the cloud of water droplets was replaced by tiny ice crystals! Further experimentation showed that there was nothing specific about the dry ice effect other than the extremely low temperature which it provided. To validate the point, he performed an experiment in which the box was at its previous temperature (below 0° C but nothing approaching the temperature of dry ice) and into it he inserted a needle which had previously been cooled in dry ice. The result was the same as if the whole unit had been cooled with dry ice. Once an ice crystal was present, it served as a nucleus for the formation of others.

With this fact firmly established, Schaefer set about the study of seeding clouds to produce rainfall. In November 1946 Langmuir observed with binoculars from the ground fourteen thousand feet below while Schaefer pushed several pounds of dry ice from an airplane into the atmosphere and had the thrill of seeing the formation of glistening snow crystals.

It is mind-boggling to realize how just a few people, working at different times and for different purposes, discovered so much about air/sea phenomena. Among the initial goals of their studies were military objectives and the solution of a health problem; factors that contributed to the results achieved were peculiar circumstances and the hobbies of several people. Langmuir and Schaefer actually enjoyed being on Mount Washington in the howling winds and freezing temperatures; Blanchard enjoyed his hobby of reading old technical writings. Elements of luck were also present in some of these events, but in noting the advances made in air/sea interactions one is struck by the fact that serendipity doesn't seem to happen to those not blessed with curiosity.

Author's thanks: This chapter was written as a summary of material supplied by Duncan Blanchard, who subsequently edited it for accuracy. I am most grateful to him for his great contribution

Chapter 4

Penicillin: Some Little-Known Factors

"I should like to point out that the possibility that penicillin could have practical use in clinical medicine did not enter our minds when we started our work on penicillin. A substance of the degree of instability [that] penicillin seemed to possess according to the published facts does not hold out much promise for practical application. If my working hypothesis had been correct and penicillin had been a protein, its practical use as a chemotherapeutic agent would have been out of the question because of anaphylactic phenomena, which inevitably would have followed its repeated use. From the scientific view, however, the problem of purifying penicillin and isolating the substrate on which I thought it acted was of interest and hence well worth pursuing."

—Ernst Boris Chain
(as quoted by Arthur Kornberg in a symposium in San Antonio, 1992)

The discovery of penicillin was accidental. Its development resulted from a succession of unlikely events, not just the one that most people have heard about. Yet it became a lifesaving treatment around the world and prompted more searches for antibiotics, which turned out to be highly important.

The Prelude to the Discovery

Before Alexander Fleming's 1928 observation that a contaminant on an agar plate had destroyed a bacterial culture, there had to be an awareness on his part that such an occurrence was within the realm of possibility. In the long quest for the "magic bullet," a succession of failures had beset many researchers, but Fleming was the beneficiary of an accidental discovery in 1921 that may have conditioned him for the event to follow. He was prone to having colds, and in London's wet climate it was not unusual for him to have a runny nose. One day he was examining an agar plate when a nasal drip fell onto the plate he was examining. That is the epitome of poor technique, but he noticed that in the area where the drip fell, there was a considerable lysis (dissolution) of the bacterial culture growing on the plate.

This was so unusual that Fleming repeated the experiment, and he was delighted to find that he could duplicate the result. He then expanded the study to include the effect of tears and found that they also caused the lysis of bacteria. Eventually he established that the active component in both the nasal drip and the tears was an enzyme that was given the name "lysozyme."

The Prelude to Ultimate Success

In the interest of clarity, this narrative will not be in chronological order. Instead, we will begin with discussions of some of the principal figures in the development of penicillin, the most important being Howard Florey, whose own health problem attracted him to Fleming's work. Florey suffered from abdominal distress, causing him to consult with Fleming in January 1929. He wondered whether lysozyme might alleviate a condition caused by an excess of mucus in the stomach.

In May of that same year, Fleming submitted his article on penicillin to the *British Journal of Experimental Biology,* of which Florey was an editor. Several years later in 1935, Florey was appointed to the prestigious Chair of Pathology at Oxford, with a mandate to invigorate the department that had deteriorated during the long tenure of his predecessor. Under Florey, the department became engaged in a study of lysozyme, and their literature search brought them into contact with Fleming's 1929 report on penicillin.

The Discovery

In 1928 Fleming was invited to write a section of a book entitled *A System of Bacteriology* because of his expertise in that field. *Staphylococci* were his specialty and he made cultures of these bacteria from diverse sources—boils, carbuncles, and other infections. According to the account by Gwyn Macfarlane in *The Man and the Myth*, Fleming had presented a paper describing variations in the color of *Staphylococcus* colonies that seemed to be related to their virulence.

Fleming returned from a vacation in early September and was rummaging through the various petri plates in his laboratory to see what growths had developed in his absence. Fleming had tossed one of the contaminated plates into a shallow dish containing Lysol before leaving on his vacation, but the dish was so full of plates that the culture never made contact with the Lysol! On the plate in question there had been a typical growth of *Staphylococcus* but there was also a contaminant, a fungus. What was interesting was that for a small distance around the fungus colony there was a zone in which the *Staphylococcus* had been lysed. Now, it was well-known that the compounds formed by a microorganism on an agar culture can migrate through the agar, and here was a case where some compound formed by the fungus was effectively dissolving the bacteria on the surface.

Fleming was aware of the potential importance of this chance event and showed the plate to several others, who were evidently not impressed. Among those who saw the plate was his former collaborator, D. M. Pryce. Another was C. J. LaTouche, whose microbiological laboratory was one floor below Fleming's at St. Mary's Hospital in the Paddington section of London. Fleming asked LaTouche to identify the contaminant, and his tentative finding was that it was *Penicillium rubrum.*

Slow Progress

Fleming's account of the event was simply that the contaminant had lysed the *Staphylococcus* cells. When he tried to repeat the experiment, however, he was unsuccessful. Both the fungus and the *Staphylococcus* grew, but they had no effect on each other. Others tried to reproduce Fleming's epic discovery and were unable to do so. Undoubtedly this was one of the reasons that Fleming encountered such difficulty in getting the medical profession to endorse penicillin. It was not until many years later that an explanation for this incongruous situation was developed by Ronald Hare, whose work is described later in this chapter.

Boris Chain had access to a culture of *Penicillium notatum* (LaTouche's initial identification of the *rubrum* was incorrect) through Margaret Campbell-Renton, whose laboratory was just down the hall from Fleming. She had collaborated with Fleming in a project and had maintained the culture in a viable condition. Therefore, Chain was able to work with the organism Fleming had found though he had not been in touch with Fleming. Chain made a major contribution to the development of penicillin by using a freeze-drying process to concentrate the compound. With this concentrated form Florey and Chain tested the effect of penicillin on three sets of mice infected with *Staphylococcus, Streptococcus,* and *Clostridium septicus.* The results, to use Andre Maurois's phrase, "smacked of the miraculous" and were published in a note in *Lancet* in August 1940. This was all news to Fleming, because he didn't even know the work was taking place. When Fleming went to visit the successful researchers, Chain was amazed to meet him because he thought that Fleming was dead!

So Fleming was not part of the team that had performed the invaluable dual task of showing that penicillin was not toxic to animals and that it cured bacterial infections that often were lethal. With the success achieved by Florey and Chain, Fleming had a renewed interest in penicillin. The preparation they had given to him was far more effective than the dilute preparation he had experimented with many years before. Also, it was more effective than the new sulfonamide drugs.

Production

Early efforts to produce penicillin were unsuccessful. In November 1940, Fleming received a letter from a friend at Parke-Davis and Company with the request that he ask the Oxford workers if they would send *Penicillium* cultures and extracts to them for possible commercial development. Fleming's response was that the Oxford team was using his original culture, therefore Fleming simply sent his own culture to Parke, Davis without mentioning it to the Oxford team. Parke, Davis set up a production process, but many technical difficulties arose, and the project was abandoned. Another attempt to produce penicillin from a culture supplied by Fleming was an effort headed by M. H. Dawson of Columbia University and the Presbyterian Hospital in New York. Evidently that effort was unsuccessful also. Failures were recorded, too, by the Lederle Laboratories.

From a medical standpoint, penicillin showed tremendous promise, but its manufacture was a long way off. Florey tried to raise funds for its development in Britain but was largely unsuccessful. By this time Britain was on the verge of war, and the need for effective treatment of wounds was the greatest ever. (Chain was quoted, however, that his motivation had been merely a scientific interest in the problem.) There had been some activity by the Wellcome Corporation, but they had gotten nowhere with the project. It was in this vacuum of possibilities that Florey decided that the William Dunn School of Pathology at Oxford should undertake the enormous task of producing penicillin. Estimates were made that if they produced five hundred liters of culture filtrate per week, within a few months there would be enough penicillin to treat five or six patients. Norman Heatley was put in charge of the production process.

In January 1941 the team made a more pure product and had enough on hand to carry out a clinical trial. This purification procedure was made possible by Chain's idea of using freeze-drying. It was still not known to Florey and Chain whether penicillin was harmful to humans, although Dawson's work at Columbia University had already shown that it was not. In their experiments, Florey and Chain had used rats, mice, and cats with excellent results. It was fortunate that they did not use guinea pigs, because subsequent tests showed that penicillin is toxic to guinea pigs. As an extension of the potential problem with human toxicity, it was decided to try the antibiotic on a person with an incurable disease who had nothing to lose and everything to gain. The subject chosen was a woman suffering from cancer. Although she eventually died, it was clear that penicillin was not toxic to her. Next was an attempt to save the life of a policeman who was dying from an infection that had progressed for two months. His condition improved dramatically, but unfortunately there was not enough penicillin available to save him. Other cases followed in which clinical successes were achieved, even though in some instances the patients died of other causes.

In June 1941 Florey turned to the United States for help. The Rockefeller Foundation in New York City agreed to pay expenses for Florey and Heatley to visit this country for an extended period. Their first success was in a meeting with Charles Thom of the United States Department of Agriculture, who understood the great promise of penicillin but also had an appreciation of the enormous task to be accomplished. Florey and Heatley calculated that it would take two thousand liters (more than five hundred gallons) of culture to produce enough penicillin to treat one severe case of infection. Huge facilities would be required, and the most likely site was the large, and brand-new USDA fermentation laboratory at Peoria, Illinois. This proved to be an excellent choice, because not only were the facilities ideal, the timing could not have been better.

There had been a bumper corn crop in Illinois that year and, in searching for uses of the corn, USDA had produced large quantities of what was termed corn steep liquor. The hope was that some use could be found for this product and thereby provide extra income for the farmers. Because of these concurrent conditions (the need for a way to produce penicillin, and an abundance of a material that might be useful), the USDA staff wanted to see whether the corn steep liquor might increase the rate of penicillin production. Florey and Heatley were content with the low yield (0.0001%) of penicillin they had been getting and were reluctant to change it. The USDA was insistent, however, that a trial be made with corn steep liquor in the penicillin production process, and it turned out to be successful. Andrew Jackson Moyer, a mycologist at Peoria, substituted lactose for sucrose in the growth medium, which resulted in a significant increase in penicillin. Then he incorporated corn steep liquor into the medium and that resulted in a tenfold increase. Whereas Florey's requirement of the mold juice for treating one infection was two thousand five hundred liters, the projected new requirement was only twenty-five liters. For his work Moyer was inducted into the National Inventors' Hall of Fame in 1987.

During this period of searching for ways to increase penicillin production, one of the workers at Peoria picked up a rotting cantaloupe and brought it back to the laboratory to determine whether the organism responsible for the rotting might produce penicillin. Dr. Kenneth Raper was the research leader in charge of the culture collection in the laboratory, and he identified the organism covering the rotting cantaloupe as *Penicillium chrysogenum.* Fortunately, this species produced penicillin at a much faster rate than *P. notatum,* and the combination of its use and that of the submerged culture made the process work so well that penicillin could be made in quantities large enough to save countless lives. Many culture improvements were made through the years. Titers of approximately 125 penicillin units per ml were increased to 250, 500, 1,000 units per ml, and by 1970 the industrial penicillin yield was 20,000 units per ml.

Another relevant finding was made by A. D. Gardner, working with Howard Florey in 1940, when he observed through his microscope the performance of bacterial cells in the presence of penicillin. Old cells were not affected, but when they divided and produced new cells, the young cells became swollen and then burst. In retrospect, this explained an earlier observation by Fleming that there was a delay in the action of penicillin on *Staphylococcus.*

Investigations by Ronald Hare on the sequence of events that must have taken place for penicillin to be discovered were painstaking and fascinating. He felt that Fleming must have inoculated the plate with *Staphylococcus* but did not incubate it and left it out on the bench where it became contaminated with the fungus. If the culture plate remained cool for several days, the fungus would grow at its normal rate (maybe in the midteen Centigrade temperature range), but the *Staphylococcus* would grow at a reduced rate. Then, if there were a hot spell, the *Staphylococcus* would grow more rapidly until it neared the zone where the metabolites of the fungus had spread. Therefore, an emerging growth of new cells in the area of the culture where the *Penicillium* metabolites were present would be inhibited.

Weather records for London at that time confirmed Hare's theory; there was a heat wave in mid-July but then a cold spell from July 27 to August 6 with maximum temperatures between 16° and 20° C. It was during this cold period that Fleming had been on vacation. After that the weather was warmer with temperatures up to around 26° C. If Fleming had begun this study during the cold spell, the fungus would have grown first at the low temperature and then the *Staphylococcus* at the higher temperature, but the newly formed penicillin would deter the formation of new *Staphylococcus* cells.

The Sequence of Improbable Events

Because of its complexity, it is easy to lose track of the elements of the penicillin story. In summary, therefore, here is the sequence of events which led from the initial discovery to the final success:

- A drop of nasal mucus fell onto an agar plate in 1921, resulting in the lysis of the bacteria on the plate. This alerted Fleming to the possibility that bacteria might be controlled by a naturally occurring compound.
- In 1928 a fungus contaminant fell onto a plate containing *Staphylococcus.* That plate was thrown into a shallow dish of Lysol to kill the organisms present
- Because the Lysol bath was so full of other plates, the bacteria and the contaminant on the historical plate were never touched by the Lysol.
- While Fleming was on vacation, there was an unusual sequence of temperatures that fostered the growth of the contaminant in such a way that it was toxic to *Staphylococcus.*
- Florey had become aware of Fleming and what turned out to be penicillin because he was one of the editors of the *British Journal of Experimental Pathology*
- Florey's personal stomach problems had caused him to have an interest in lysozyme, which had been discovered by Fleming.
- Lysozyme became the focus of Florey's attention when he took over the chair of pathology at Oxford. He hired Boris Chain to work on the project, and ultimately Chain became involved with the penicillin project. His use of the freeze-drying process resulted in a highly increased concentration of penicillin.

- It was fortunate that Chain and Florey used mice and rats to test penicillin. Had they used guinea pigs as the test organisms, the guinea pigs would have died and the conclusion likely would have been that penicillin was too toxic to be considered for human use.

- Because of their need for large fermentation facilities, Florey and Heatley were directed to the new USDA laboratory at Peoria, Illinois, and at that time USDA was hoping to find a use for what it termed corn steep liquor. It turned out that corn steep liquor increased the yield of penicillin.

- At Peoria a worker picked up a rotting cantaloupe at a fruit market, and the strain of *Penicillium* on that cantaloupe was a prolific producer of penicillin.

So a succession of unlikely events led ultimately to the successful production of penicillin. Fleming had given up on the project after eight years of futile relationships with the medical community and even some of his colleagues. Because of Florey, Heatley, and Chain, the world has benefited from penicillin, and its potency led others to search for new antibiotics.

* * * *

To make this chapter easily readable there have been no insertions of footnotes that inevitably divert attention from the text. In fact, however, all of the material in this chapter can be traced to the sources shown below:

Kauffman, George. 1978. The penicillin project from petri dish to fermentation vat. *Chemistry*. September, 11–17.

Macfarlane, Gwyn. *Alexander Fleming, the Man and the Myth*. Cambridge, MA: Harvard University Press, 1984.

Maurois, Andre. *The Life of Sir Alexander Fleming, Discoverer of Penicillin*. New York: Dutton Press, 1959.

CHAPTER 5

AN IMPROBABLE ROUTE TO A CURE FOR TUBERCULOSIS

In 1910 Selman Waksman emigrated to this country from Russia at the age of twenty-two and received the Nobel Prize in medicine or physiology in 1952 for the discovery of streptomycin. It proved to be a successful treatment for tuberculosis, though some had thought there never would be a cure for TB because *Tubercle bacilli* live in inaccessible places in the lungs.

Many elements of serendipity and chance can be found in the story of streptomycin. The pages that follow are essentially a condensation of the excellent narrative written by Julius Comroe Jr. which begins with 1914 when antibiotics were unknown.[1] After many years of disparate studies, streptomycin became recognized as a weapon to be used against tuberculosis, but it also became apparent that an adjunct compound was also required for it to afford reliable effectiveness. Two compounds were developed toward that end: isoniazid and para-amino salicylic acid. Isoniazid was an intermediate compound that was needed for the synthesis of a compound that was considered promising; the sought-after compound turned out to be ineffective but isoniazid worked. Para-amino salicylic acid was designed as the result of what is known as a vitamin/vitamer phenomenon, and that did not come about until the appearance of the sulfa drugs in the mid-1930s.

So, in order for there to be an effective treatment for tuberculosis, there was a need for streptomycin, isoniazid, and para-amino salicylic acid. How that came about is a fascinating tale that involved numerous people.

BACKGROUND

A Correct Hypothesis

Waksman was a microbiologist whose interests were far removed from medicine; his thoughts were oriented toward an understanding of the interactions among microorganisms in the soil. However, a question he posed did have an important link with medicine: are there microorganisms in the soil that can metabolize Type III pneumococcus cells? His reasoning was that there had to be such microorganisms present unless some factor was destroying them. He decided to search for soil microorganisms that could metabolize the polysaccharide coating surrounding pneumococcus cells. Around the year 1914 he began that search by making a special agar whose only carbon source was polysaccharides of the type found in pneumococci. He then inoculated the agar with a water suspension of soil, and for any soil-bound microorganism to grow on this unique agar, it would have to be able to metabolize the polysaccharides therein.

Waksman's quest for such a microorganism was successful. Because he was at Rutgers University in New Jersey (at New Brunswick) he used soils in that area and found what he was looking for in a cranberry bog. It was identified as the fungus *Actinomyces griseus.* At the time of this event, there was no thought about its medical significance, but it was interesting to Waksman merely as an example of one soil microorganism having a property that could destroy another one.

Between 1914 and 1921 he had done research on soil actinomycetes, evaluated a synthetic chemotherapeutic agent, and had looked into the possibility of using microbes to produce important enzymes. He was aware of antagonisms between actinomycetes and bacteria because in 1941 he wrote a review of fourteen articles published between 1890 and 1935 that told of such phenomena. From 1924 to 1927 one of his graduate students, Rene Dubos, worked on the decomposition of cellulose by bacteria. Also Waksman knew of the work done by Avery and Dubos in 1930 concerning the ability of bacteria to degrade the polysaccharide surrounding the Type III pneumococcus.

Ignorance of Antibiotics

Today we take for granted the multitude of drugs available for the control of disease, but at the time of Waxman's early work the dawn of antibiotics had not pierced the horizon. There had been a hope for finding a "magic bullet," (i.e., a compound that would kill germs but not the patient). An idea of the futility of such a search was evident in 1909 and 1910 when Paul Ehrlich had tested 605 compounds before finding one, Salvarsan, that was effective against spirochetes (for example, syphilis). Only a few organic compounds were known, and finding one with medicinal properties would be time-consuming and perhaps hopeless. To realize the extent of this void of scientific knowledge, consider that Waksman wrote a 360-page book, *Enzymes,* in 1926 that devoted only one paragraph to the antagonisms between bacteria. He wrote two monographs, in 1931 and 1932, and did not even mention Alexander Fleming or his work with penicillin. In fact, he wrote a large text book (894 pages, second edition in 1932) on soil microbiology that contained only two pages on the subject "Antagonism and Symbiosis Among Microorganisms." A study had been published in 1913 by Vandremer that should have been of real interest, but Waksman may not have been aware of its existence or its significance. Vandremer had used a "macerate" of the fungus *Aspergillus fumigatus* to treat more than two hundred patients who had tuberculosis, and some had made remarkable recoveries.

Whether anyone was impressed with Vandremer's work is hard to tell, but there would have been a general bias against such an approach anyway. One of the savants of the day, Sir Almoth Wright, thought that the best protection against disease was to build up the well-being of the patient rather than to try to kill the organism causing the disease. In this atmosphere of simplicity concerning medical matters, a suggestion that a soil microorganism be used in the control of disease would probably be laughed at.

Some Missed Opportunities

In 1932 the National Research Council and the National Tuberculosis Association gave Waksman a grant of three thousand five hundred dollars for a study, "Research of Pathogenic Bacteria in Relation to Soil Populations." Waksman turned the project over to a graduate student, Chester Rhines. The work was finished by the end of 1934, and several hundred dollars were still unexpended. Rhines published three papers on the subject in 1935 and made the observation that under certain circumstances tubercle bacilli continued to grow in close association with soil microorganisms. What should have excited him, in retrospect, was that in some instances the bacilli diminished remarkably. Neither Rhines nor Waksman continued the study. Years later, Waksman philosophized on this missed opportunity by saying that pathogenic microbes were out of his field of investigation at that time. In his brilliant commentary on the subject, Julius Comroe faults the two committees that funded the study for not following up on this promising lead.

One of the players in the streptomycin saga was Fred Beaudette, a participant in the discovery of the Newcastle disease virus. Beaudette was the poultry pathologist at the New Jersey Agricultural Experiment Station at Rutgers. About 1935 he brought to Waksman's attention an agar tube that had been covered with a healthy tubercle bacillus which had been killed by a fungus that accidentally contaminated the tube. This should have been of great interest, but Waksman evidently did not recognize its significance.

Waksman's own son Byron (a student of Julius Comroe) wrote to his father in 1942 suggesting an idea for a summer project. Byron was impressed with the simplicity of the method his father had used in detecting antibiotic factors such as the one that led to the discovery of *Actinomyces griseus* more than twenty-five years earlier. Byron suggested that the same method be used to isolate a number of strains of fungi that would act against *Mycobacterium tuberculosis.* Perhaps the elder Waksman was too distracted by other affairs, but he wrote back to Byron that he didn't think the time had come for that. Byron's idea would have been a direct approach to the problem of finding a cure for tuberculosis and was worthy of a major program.

PROGRESS

One of Waksman's graduate students at Rutgers was Rene Dubos, who ultimately became renowned for his work at the Rockefeller Institute for Medical Research. A study that he conducted in the late 1920s was patterned after the approach Waksman had used in discovering *Actinomyces griseus* in 1915. Dubos embarked on a different approach in 1939 when he used spore-forming bacteria to "unclothe" Type III pneumococcus cells; the human body is capable of destroying these cells once their coating has been removed by a process known as phagocytosis. Dubos's work reawakened Waksman's interest in the subject, and he proceeded to build on what Dubos had done. In the late 1930s Waksman enriched the environment of bacteria in the soil by incorporating gram-positive bacteria into it. This induced another specific antibacterial enzyme to kill gram-positive cocci. Better yet, he was able to find a soluble lethal factor in the soil bacteria called gramicidin.

In 1940 Waksman discovered actinomycin but it was too toxic for use on humans.

In 1943 Beaudette came again to Waksman, this time with a tube containing an agar that he had streaked with a swab from the throat of a sick chicken. Waksman gave the culture to an assistant for identification and it was *Streptomyces griseus,* the same organism known earlier as *Actinomyces griseus.* Waksman was interested in Beaudette's culture, and the response was described this way in Comroe's account:

> The difference between 1915 and 1943 was that Waksman was now screening soil organisms for antibiotic activity. *S. griseus* happened to produce streptomycin, and streptomycin turned out to be an antibiotic with broad activity against bacteria, including the tubercle bacillus. Was the organism that Waksman grew in 1915 a streptomycin producer, or a slightly different strain? No one will ever know for sure, but microbiologists have a habit of preserving over many decades a specific strain of an organism by the process of regularly rejuvenating it by transferring it onto fresh media. Waksman had done this for his original 1915 *Actinomyces griseus.* In 1949 Kelner found that it did not produce streptomycin spontaneously, but did so after being irradiated. It is possible that the 1915 culture was a busy streptomycin-producer in 1915, lost this property because of repeated cultivation on artificial media, and regained it in 1949 when irradiation caused mutation.

There are divergent views on the importance of the culture derived from Beaudette's sick chicken. Evidently the *Streptomycin griseus* isolated from the chicken was effective in the laboratory tests with the tubercle bacilli in 1943. But there is also good reason to think that the organism was taken from a soil picked up by Waksman's assistant, Albert Schatz. In 1994 Karl Maramorosch of Rutgers published a letter in which he claimed the primary credit was due to Schatz, pointing out that streptomycin was first reported in 1944 in "Proceedings of the Society for Experimental Biology and Medicine."[2] That article, "Streptomycin, a Substance Exhibiting Antibiotic Activity Against Gram-Positive and Gram-Negative Bacteria," was authored by A. Schatz, E. Bugie, and S. A. Waksman, in that order. The two papers that followed were by Schatz and Waksman. The patent for streptomycin was issued to Schatz and Waksman—not to Waksman alone nor to Waksman and Schatz.

For our purposes the substantive fact is that a major advance had been made in combating tuberculosis, and it began with a study oriented toward the interactions of microorganisms in soil. Whether the principal credit many decades later was due to Waksman or to Schatz might still be a subject for debate.

TWO NECESSARY PEOPLE

Two other persons played essential roles in the streptomycin saga.[3] W. H. Feldman and H. C. Hinshaw discovered that streptomycin slowed, stopped, or reversed human pulmonary tuberculosis. Feldman came to the United States from Scotland in 1894 at the age of two, became a veterinarian and ultimately a pathologist at the Division of Experimental Medicine at the Mayo Clinic. He then was on the staff of the Department of Comparative Pathology in Mayo's graduate school. Hinshaw's background was in zoology, parasitology, and bacteriology before he was awarded his medical degree at the University of Pennsylvania. He taught Comroe at Penn, which explains Comroe's encyclopedic grasp of the streptomycin story. When a position opened up at Mayo in the Department of Pulmonary Diseases, Hinshaw became interested in pneumonia and its treatment with the new chemotherapeutic agents. Between them, Feldman and Hinshaw had experience with a broad spectrum of diseases and victims, both animals and humans. Each was dedicated in a singular fashion to finding a cure for tuberculosis.

Feldman and Hinshaw maintained that tests being designed for the assessment of new drugs, particularly for tuberculosis, had to be more rigorous. Feldman knew that Waksman was screening actinomyces for possible antibiotic activity and visited his laboratory in November 1943. Comroe points out that Feldman urged Waksman to include tubercle bacilli in his studies, and further suggested that he use a virulent strain from human TB rather than what might be a worn-out strain from the student bacteriology laboratory. He also urged that sufficient time be allowed for the tubercle bacillus to be well established in a test animal before the administration of the streptomycin. Some protocols had called for the inoculation with tubercle bacillus and an antibiotic at the same time.

Feldman asked Waksman to let him know when an anti-tubercle bacillus compound showed up. The following March (1944) Waksman obliged by writing to him and asking whether he was prepared to test streptomycin on guinea pigs.

A ten-gram supply of the drug was used by Feldman on four tubercular guinea pigs, but the supply ran out in June 1944. In a meeting with Waksman and representatives of the Merck Company, it was agreed that Merck would supply more streptomycin if Feldman and Hinshaw would also include strepthricin (a potential Merck product that had been supplied by Waksman) in the tests with tubercular guinea pigs. Beginning in July 1944, they repeated their pilot study with a regimen that was extremely taxing to Feldman and Hinshaw. They had inoculated many guinea pigs with the tubercle bacillus and over a period of forty-nine days the infections of these test animals had become severe. By mid-August they were ready with twenty-four control subjects, twenty-five animals with virulent TB, and a new batch of streptomycin from Merck. Without technicians to help them, Hinshaw and Feldman administered streptomycin to all subjects once every six hours for 166 days before concluding that the test was a great success. Counting the forty-nine days required to develop a virulent strain in the guinea pigs, the whole test required 215 days.

Comroe provides the following account of the first human treated with streptomycin:

> The first patient treated with streptomycin was a twenty-one-year-old woman in the last stages of tuberculosis. She was in the Mineral Springs Sanatorium at Cannon Falls, Minnesota, under the care of Dr. Karl Pfuetze, who was collaborating with Feldman and Hinshaw. Treatment began on November 20, 1944, fifteen months after Waksman first isolated *Streptomyces griseus.* Between then and April 7, 1945, she received five courses of treatments, each lasting ten to eighteen days. She improved markedly, and two years later her case was closed as "apparently arrested pulmonary tuberculosis."

Separate studies followed. By September 1945 they had treated thirty-four tuberculous human patients and reported encouraging results. Feldman and Hinshaw were not publicity seekers. They were careful in discussing the subject to make no claims except those that were verifiable. They shunned reporters, preferring to inform only their scientific compatriots with formal articles and interviews with those having technical training. Streptomycin was still not recognized as a cure for TB.

Adjuncts to the Story

Streptomycin had unfortunate side effects (mainly a potential loss of hearing) besides being not totally satisfactory as a cure for tuberculosis. Evidence had been established that the tubercle bacilli could develop a resistance to streptomycin. It was generally agreed that a second antituberculosis drug was highly desirable, and two were developed. In the accounts below, the details of the development of para-amino salicylic acid, beginning in 1939, and isoniazid in 1951, are given. Both are effective treatments when combined with streptomycin.

According to an account by George Kauffman (the best-known historian of chemistry in modern times), the pharmaceutical firm of E. R. Squibb and Sons had developed a procedure for *in vivo* testing of compounds for antituberculosis activity.[4] A team of twenty-four researchers there had tested more than eight thousand compounds toward this end. The drug used as a standard to evaluate this procedure was called tibione; it was an active drug but had bad side effects.

In 1939 Jorgen Lehmann, a pharmacologist and cell biologist in Goteborg, Sweden, was aware that the onset of World War II would cause a proliferation of tuberculosis, and he had an interest in potential cures for the disease. Lehman had read a short note by F. Bernheim to the effect that bacilli doubled their oxygen uptake when provided with sodium salicylate. Lehmann reasoned that if the bacilli required sodium salicylate as a nutrient, perhaps some slight modification in its structure would interfere with the metabolism of the bacilli. There was precedence for this approach, because it was about that time that D. D. Woods had correctly speculated on the mechanism by which sulfa drugs were effective. In what is called a vitamin/vitamer relationship, one compound is a nutrient, but another compound with a slightly different chemical structure is poison. For bacteria para-amino benzoic acid was a nutrient, and sulfanilamide was the poison. With this concept as a template, Lehmann proposed that if sodium salicylate were a nutrient, the para-amino salicylate (PAS) might be a poison of the tubercle bacilli. Lehmann had no way of preparing PAS, so he wrote to the pharmaceutical company Ferrosan in Malmo, Sweden, and asked that they prepare it as well as several close isomers. Ferrosan had difficulty preparing PAS, but on Nov. 11, 1943, sent five grams of the compound to Lehmann. The first tests against the tubercle bacillus were failures. They had been conducted in the standard manner in which the antibiotic was applied at the same time as the bacillus was inoculated. When Lehmann had heard that better results had been obtained when the antibiotic was applied only after the bacillus had been given a head start, he tried it in that mode and was successful. On December 3 Merck sent an additional thirteen grams. By December 30 it had been shown that PAS inhibited the growth of the tubercle bacillus at a concentration lower than that reported for any previous compound.

In 1951 researcher Harry L. Yale designed a compound that bore a structural resemblance to tibione, but the synthesis for this new compound required that isoniazid be prepared as an intermediate. Squibb's policy at that time was to submit all new compounds to screening tests of various sorts, and for that reason isoniazid was included among the compounds being tested. It turned out to be more effective against tuberculosis than any other compound, and it had minimal side effects. Isoniazid then became one of the compounds to be associated on a regular basis with streptomycin.

So now there were two compounds available as adjuncts for use with streptomycin for the treatment of innumerable patients.

Sequels, Good and Bad

Beginning in 1946 three large trials with streptomycin gave promising results but also indicated that streptomycin alone was not sufficiently effective. Clinical trials with PAS in Sweden were reported in 1950 with very good results. Not long after that it was apparent that the combination of streptomycin with PAS (or several other compounds) was required for the control of tuberculosis.

Twenty eight years is a long gestation period for the development of a drug. But the TB sanatoriums are now only a distant memory, and that is cause for rejoicing. Streptomycin, PAS, and isoniazid were not in anyone's thoughts at the beginning of this saga, but chance, luck, and perseverance combined to memorialize them.

REFERENCES

1. Comroe, Julius Jr. 1978. Pay dirt: The story of streptomycin. Part 1. From Waksman to Waksman. *Amer. Rev. of Resp. Diseases.* 117: 773.

2. Maramorosch, Karl. 1994. Letter to the editor. *The Sciences.* Jan/Feb. Pg 6–7. 1994.

3. Comroe, Julius Jr. 1978. Pay dirt II. The story of streptomycin. Part II. Feldman and Hinshaw; Lehman. *Amer. Rev. of Resp. Diseases.* 117: 957–968

4. Kauffman, George B. 1978. The penicillin project: From petri dish to fermentation vat. *Chemistry.* September: 11–17

CHAPTER 6

CONTRIBUTIONS OF ASPIRIN, RAT POISONS, AND OTHER COMPOUNDS TO HEALTH

"The solution of a biological problem when worked out often seems so reasonable and so simple that wonder is expressed that any other sequence of events could be contemplated, but those who think along these lines can have had little experience in opening up new fields of biological inquiry."

—John Mellanby[1]

Mellanby uses the term "biological," but his words hold true also for medical or physiological problems. Just to make a start toward an understanding of the human body was an enormous task. A great advance was made by Sir William Harvey in 1616 when he proved the existence of the circulatory system, and in that instance he was helped by the physical evidence from countless observations of animal systems. How, though, does one begin to know that there are glands that perform unique functions, or that there is a system for regulating the body temperature? Before the advent of organic chemistry, how would one know that there were amino acids, vitamins, and proteins involved in nutrition? Progress could only be made slowly, and luck would be vital to the process.

DEVELOPMENTS IN DRUGS

We are now accustomed to CAT scans, laser surgery, mammograms, highly specific drugs, and technological innovations that are invaluable in diagnosing medical problems or treating them. Today's young adult might find it hard to believe that it is only relatively recently that antibiotics became available. To illustrate the point, a little personal history might be useful.

In 1936 two crises in my family would have been eased considerably had there been medications that we take for granted today. First, I developed a severe case of trench mouth that resulted in such swollen gums that my teeth were hardly visible. Gentian violet (potassium permanganate) was the treatment that was prescribed, but its only apparent effect was to discolor my mouth for an extended period. Subsequent tooth decay was so extensive that I probably spent as much time in the dentist's chair as anyone. The second crisis, much more serious, involved my brother Frank, who suffered from a ruptured appendix. That led ultimately to peritonitis, which nearly took his life. Had the wide range of antibiotics now available been known in 1936, probably neither Frank nor I would have had any problem.

Aspirin: Its Origin

Aspirin is known universally as a pain-killer, and in recent years other uses have been found for it. Statistics concerning aspirin boggle the mind. In 1996 there were twenty-nine billion aspirin tablets produced in the United States and fifty billion sold worldwide (data provided by the Bayer Company). Americans use about one hundred ten tablets per person per year. The history of aspirin goes back to Hippocrates two thousand four hundred years ago, who recommended chewing willow leaves for relieving pain during childbirth and applying the sap of the poplar tree for eye diseases. Events in the 1600s and succeeding centuries identified the salicylates as being beneficial in the treatment of fevers and pain. Synthetic production of salicylates began in Germany, due largely to the practical method of preparing salicylic acid from phenol. This development was attributed to Kolbe and Lauteman at Marburg University, after which Kolbe joined the Bayer Company. In the pharmaceutical division of Bayer was a young chemist, Felix Hoffman, who was assigned the task of developing a salicylic acid derivative with maximum analgesic powers. Hoffman had an extra motive to succeed, because his father suffered intense pain from arthritis. In 1897 Hoffman synthesized acetylsalicylic acid and after ingesting some to determine its effect, administered it to his father. It was the most effective pain reliever that the elder Hoffman had used, and the fame of what came to be known as aspirin spread widely.

The anti-blood-clotting effects of aspirin did not become apparent until 1940, following a succession of discoveries of other blood-thinning compounds (heparin, cephalin, dicumarol, and warfarin). Unusual circumstances played a role in each discovery.[2]

Understanding Blood Thinning

To perform its most important bodily function, blood must be a liquid. There are times when it must form a clot, and at other times that clot can have disastrous consequences. Obviously an understanding of both processes is needed.

Heparin tends to prevent clot formation, but its discovery originated with a clotting agent, and its discovery depended on one person.[3] We start with thromboplastin, an enzyme occurring in blood platelets that functions as a clotting agent. In 1912 the world's most prominent expert in coagulation was William Howell of Johns Hopkins University, who was studying the pro-coagulation effects of a thromboplastin extract he called cephalin.

Onto the scene came Jay McLean of San Francisco, a medical student at the University of California, who had an intense interest in blood clotting. McLean wanted to transfer to Johns Hopkins but was refused admission. Despite this rebuff, he left California, worked at menial jobs as he made his way east, and knocked on the door of the admissions office at Johns Hopkins. That office had already made its judgment about young McLean but agreed to his offer to work in a research laboratory for a year while awaiting admission to the medical school. McLean then called on William Howell to tell him that his goal was an academic career in surgery, and that he wanted to work in physiological research in Howell's laboratory. Accordingly, Howell assigned to him the task of isolating the active pro-coagulant in cephalin.

Cephalin is a mixture of phospholipids, and in seeking a pure coagulant McLean extracted phospholipids from the dog heart and liver. Early in 1916 he found that these extracts, particularly from the liver, lost their clotting activity with the passage of time and became anticlotting instead. When McLean told Howell that he had found an anticlotting agent, Howell was skeptical. However, he became convinced when McLean placed a beaker of blood on his desk, added some "heparphosphatide" to it, and showed that the blood didn't clot. This prompted Howell to drop all other pursuits and actively study the anticlotting activity of this extract.

At about this time, McLean described his discovery at a meeting of the Society of Normal and Pathological Physiology at the University of Pennsylvania and wrote his only paper on the subject that same year. Howell opposed McLean's actions, prompting McLean to leave Johns Hopkins and accept a research fellowship at the University of Pennsylvania the following year. McLean dropped the work on the anticlotting agent and devoted himself to the practice of surgery. He never went into the academic field he had sought initially.

Howell named the active ingredient "heparin." With the collaboration of E. Holt in 1918, he found heparin in other organs and confirmed its solubility in organic solvents. In 1922 Howell reported finding an anticoagulant in the dog liver that was water soluble. This had to be different from McLean's phospholipid and the "heparin" he had worked with in 1918. Between 1922 and 1928 Howell and Holt found that the new heparin was not a lipid of any kind but rather a sulfated carbohydrate present in peptone shock blood, which would explain its solubility in water.

Neither McLean nor Howell tried to patent heparin, and Howell gave the preparation of the compound without charge to a Baltimore pharmaceutical firm. The company was not able to produce the dog-liver heparin with sufficient purity. Subsequently J. Erik Jorpes at the Karolinska Institute in Stockholm produced a highly pure heparin and determined its chemical structure. The year was 1935, about twenty years after Jay McLean first became involved in the problem. For the following several years, heparin passed all the necessary tests for toxicity and efficacy and became recognized as being effective in preventing thrombosis.

The next step was to determine whether it would be effective in treating patients who already had problems with thrombosis. The first patient in the United States to be treated for thrombotic problems was Arthur Schulte, a young man with malignant recurrent thrombophlebitis. His physician was Irving Wright, whose personal suffering with thrombophlebitis led to his study of anticlotting compounds. Schulte came to Wright in late 1938, and Wright then contacted C. H. Best in Toronto and asked him to come to New York with enough heparin to treat Schulte. After sixteen days his supply of heparin ran out and other blood thinners had to be used. Schulte was still taking warfarin as a blood thinner and whenever he would be taken off warfarin because of bleeding, his thrombophlebitis would recur. Heparins are now being studied for their ability to enhance endothelial function, which was touched on in the work of Gruberg and

Raymond, whose work pointed to the importance of methionine in that regard. It seems as if the expanding need for knowledge in any subject never ends.

How Spoiled Sweet Clover Led to Understanding Blood Clots

The relationship between coumarol, its derivatives, and a precursor to prothrombin became known about 1932. But the story began three decades earlier.

Around the turn of the century, farmers in northern prairie states had difficulty in growing sufficient animal feed crops. The problems were overfarming and a harsh climate. Consequently they began planting a yellow-flowering sweet clover plant, melilot, that had been imported from Europe. This produced abundant silage for cattle but within several decades was the cause of a new disease that killed many cattle. Called "sweet clover disease," it caused fatal bleeding. A pathologist named Schofield in 1921 in Alberta, Canada, determined that the spoilage of sweet clover was the cause of the problem. He also showed that by removing the sweet clover from the diet of cattle and transfusing them, the disease abated.

By 1932 it was established that the affected cattle were deficient in prothrombin, a protein produced in the liver and converted into thrombin in the clotting of blood. A key point in the saga of sweet clover disease was that Karl Paul Link was offered a job in late 1932 by Ross Gortner at the University of Minnesota. Gortner wanted Link to work on the sweet clover disease project, but Link instead accepted a position at the Agricultural Experiment Station at the University of Wisconsin in Madison. The goal of this assignment was to develop a strain of sweet clover that was free of, or at least low in, coumarin. At the time, there was no suspicion that coumarin was related to sweet clover disease. Link had found that none of the more than sixty compounds they worked with was pathogenic. Then occurred one of those serendipitous events that played an important role in the history of science.

Cattle belonging to a farmer named Ed Carlson were killed by the sweet clover disease, and in a howling blizzard in February 1933, Carlson called on the experiment station for help. He was directed to Link in the Biochemistry Building, to whom he showed a dead cow, a milk can containing blood that would not clot, and one hundred pounds of spoiled sweet clover. Link's only advice was not to use the clover and to have the cows transfused. Carlson was too poor to do either and, crestfallen, drove home into the blizzard. Link felt so sorry for Carlson that he set out to find a cure for the sweet clover disease, the same chore he had turned down when offered the job at the University of Minnesota.

By 1935 a prothrombin time test (referred to as PT) had been developed by A. J. Quick and co-workers to quantify changes in blood clotting time caused by various agents. They showed that the time required for clotting was elevated in the sweet clover disease and in hemorrhagic chick disease. In June 1939 Link's associate, Harold Campbell, isolated a hemorrhagic agent from sweet clover and even isolated a pure crystal of it. In 1940 they reported on the hemorrhagic properties of the compound. A colleague, C. F. Huebner, solved the structure of the compound and synthesized it. The compound, a derivative of 4-hydroxy coumarin, was given the name dicumarol. Extensive tests showed that the addition of dicumarol to hay or sweet clover had the same effect on cattle as did spoiled sweet clover. By late 1940 dicumarol was being administered to human patients for its anticlotting properties.

A Rat Poison Used in the Treatment of a President

The concept of superthinning the blood intrigued Link. He was responsible for the discovery of warfarin and that was caused partly by his confinement to a sanitarium for tuberculosis (how remarkable it is that one health hazard led to the resolution of another one). The year was 1945 (just shortly before streptomycin had gained full acceptance for controlling TB) and Link avoided boredom by reading about the history of rodent control. Upon returning to his laboratory, he was able to show that one of the coumarin derivatives he had worked with, 3-(alpha-acetonylbenzyl) 4-hydroxy coumarin, was just what was needed. The patent was issued to a nonprofit body called WARF (Wisconsin Alumni Research Foundation), and by stretching the initials to connote warfare on rats, the name "warfarin" was given to the new compound. Because it was a therapeutic human agent, there was some unease that it was also a rat poison. One factor leading to its acceptance for treating humans was that when a Navy recruit attempted suicide by ingesting 567 milligrams of warfarin, he survived. In 1953 warfarin was tested with human volunteers in direct competition with dicumarol and was found to be superior. Even President Dwight Eisenhower was administered warfarin when he suffered a heart attack in 1955.

Aspirin: Some Connections

Link had discovered dicumarol and warfarin, both oral anticoagulants, and noted a connection with aspirin when he realized that a degradation product of dicumarol is salicylic acid. Another connection between aspirin and anticoagulation was the observation made by Lawrence Cravin, a general practitioner in Glendale, California. Cravin noticed that children whose tonsils had been removed were often given chewing gum containing aspirin to relieve the pain. Unfortunately, there seemed to be an association of the aspirin with unusual instances of bleeding. Cravin concluded that aspirin acted as a blood-thinner and might serve as a better anticlotting agent than dicumarol, which had been used for that purpose. He recommended that healthy men between the ages of forty-five and sixty-five, particularly those who were overweight or who led sedentary lives, should take an aspirin a day. Subsequently there were ten major trials of aspirin with patients who had already had some sort of attack involving reduced arterial blood flow. Different dosages were used in these trials, but in six out of ten there was an overall benefit resulting from the use of aspirin.[4]

Aspirin has touched the lives of nearly everyone. Its relationship to the action of rat poisons is probably best left unknown to the population at large.

Teatime and Interferon

In 1989 Jean Lindenmann wrote a *Citation Classic,* an informal account of his collaboration with Aleck Isaacs concerning interferons. The original paper, entitled "Virus Interference. I. The Interferon,"[5] has been cited in the literature more than one thousand times. Part of his account follows:

> I arrived at the National Institute for Medical Research as a post-doc with a Swiss fellowship in July 1956. I wanted to work in virology with Sir Christopher Andrewes. Somebody had claimed that poliovirus (at that time the real star among model viruses) could be cultivated in rabbit kidney cells, and I was assigned the task of confirming this rather important finding. This proved a disappointing experience because poliovirus just does not grow in rabbit cells. At tea time one day I was introduced to the worker next door. "This is Aleck Isaacs," I was told. He asked me what I had done before coming to England, and I replied that I had been interested in the phenomenon of viral interference; I had even done some as-yet-unpublished work on this, showing that inactivated influenza virus was capable of interfering with the growth of live influenza virus, even when stuck to the surface of red cells, from where the inactivated virus presumably could not escape. My interlocutor seemed highly interested: how did I know the virus was truly inactivated? How did I know it would not elute from red blood cells? This, I replied, was all to be read in an obscure paper by some Australians I had come across. This paper, in fact, had been written by Isaacs (author's note: the very man he was talking to) and Margaret Edney. But, when reading it, I had in my mind pronounced the

> name in mid-European fashion as something like "Ezaak," whereas the man I had been introduced to was "Eye Sacks." We immediately started a collaboration. What a relief from my frustrations with the polio work, which of course went on for several months. We repeated the work with red cells but very soon realized that of the three elements that entered into our experiments (inactivated virus, red cells as carriers of the inactivated virus, and host tissue), the red cells were an unnecessary complication and that it was the interaction of the inactivated virus with the host cells that resulted in the release of a substance, called in laboratory slang "interferon," that inhibited viral growth when applied to fresh tissue.

So that is how interferon was discovered. It began with the study of the influenza virus and ended with a product now playing a role in cancer research. A necessary connection was the chance meeting at teatime between two men who understood the value of collaboration.

An Understanding of Enzyme Kinetics

It is trite to say that necessity is the mother of invention, but that was the basis for a research article that has been referenced more than one thousand two hundred times. Wallace Cleland, of the University of Wisconsin at Madison, wrote a paper entitled "The Statistical Analysis of Enzyme Kinetic Data."[6] Cleland became an enzyme kineticist by accident when he was asked to teach a first-year graduate-level course in biochemistry. Kinetics was to be covered in the course, and to him the field was in a very unsatisfactory state. A problem for Cleland was that he never had a course in statistics:

> So when I was asked to write an article for *Advances in Enzymology and Related Areas of Molecular Biology* on the statistical analysis of enzyme kinetic data, I was placed in the position of the blind leading the blind. I believe that because of my limited knowledge of statistics I was able to express the principles in language that an ordinary biochemist could understand. Certainly no statistician I have known can write so a non-statistician can understand him.

At the risk of alienating statisticians reading this book, I think it is accurate to say that there might be general agreement with Cleland's premise. He evidently succeeded in his teaching, and at the same time guided many later researchers through the land mines of statistics.

Oxytocin: Seeing a Hormone at Work

The hypothalamus is a part of the brain involved in myriad control processes, including regulating body temperature and metabolism. A sequence of events follows the release of hormones from the hypothalamus, and Henry H. Dale discovered a relationship serendipitously when he paired one compound with a particular smooth muscle—the uterus. In *Annual Review of Pharmacology* he commented on its origins:[7]

> Nearly all the investigations in which I had the privilege to take part during the period of nearly forty years of activity which ended with my retirement, now twenty years ago, were concerned with problems or phenomena which were brought to my notice by requests, or encountered by accident while I was following such suggestions from others, and never with any plan of research which I had chosen for myself and consistently followed.

His good friend George Barger at the Wellcome Company was studying the active principles of ergot and in the mid-1930s had supplied Dale with its various components:

> In studying this action of one of Barger's ergot preparations, in paralyzing, or reversing the augmenter actions of adrenalin and of sympathetic nerves, I used a dose of post-pituitary extract as a control pressor agent, showing that its action was unaffected by the ergot-paralysis. By sheer accident, I happened to be recording in that experiment the contractions of the uterus as well, and thus saw, for the first time, the stimulant effect of the post-pituitary extract on that organ, later shown to be due to a separate hormone, oxytocin.

Subsequent experiments led to the study of the sympathomimetic activities of a series of amines, leading up to adrenaline and its immediate homologues. Dale was awarded the Nobel Prize in 1936.

Cyclosporin A and Organ Transplants

This story—not strictly an instance of serendipity—is an example of a great advance for medicine through an arduous route. The drug cyclosporin A is used during and after the transplanting of organs to prevent rejection of the new organ by the host body. How this was discovered is described below.

Switzerland's Sandoz Corporation had adopted the practice of asking its employees to bring back samples of soil from places they visited. In 1969 one of the employees on holiday scooped up a handful of Norwegian soil. He was screening organisms from the soil for their ability to inhibit other organisms, and eventually an extract of a fungus from this soil made its way to the laboratory of Jean-Francois Borel, where it would be tested for its possible effects on the immune system.

Twenty new compounds were screened per week by a procedure in which a live mouse would be treated with the test chemical, while at the same time being injected with red blood cells from sheep. Normally the sheep cells would trigger the immune system of the mouse and cause it to produce antibodies. A compound preventing the formation of antibodies was then considered to be interfering with the immune system. The soil extract used by Borel did prevent antibody formation but, as the first of a long series of setbacks, failed to do so in a confirmatory test. The Sandoz Company wanted to drop the project, figuring there wasn't much of a market for a suppressant of the immune system anyway. But it was Borel's determination to complete these experiments that led ultimately to the routine use of cyclosporin, the active ingredient of that fungal extract, in transplant operations. By the time cyclosporin AS came on the market, transplantation science was ready for this powerful immunosuppressant. Organ transplants have steadily increased in numbers as shown in the table below:

NUMBERS OF TRANSPLANT OPERATIONS[8]

Organ	1981 Transplants	1995 Transplants
Heart	62	1,952
Liver	26	3,229
Kidney	4,883	9,004
Cornea	15,500	43,743*

* Figure for 1984 at the time.

Incidentally, the reason for the failure in the second experiment with cyclosporin was that those preparing it had changed their procedure and had not told Borel of it.

Later transplant operation figures, according to the *infoplease* Daily Almanac for May 5, 2005, showed that in 2004 the heart transplants were 2,016, liver 6,167, and kidney 15,999. No data for corneas were given in that report.

The Utility of the Concept of Half-Life

A catalogue of unusual but necessary factors in the discovery process can be found in the narrative of Calvin M. Kunin, a fellow in infectious diseases under Maxwell Finland at the Thorndike Memorial Laboratory.[9] This was part of the Harvard Unit at the Boston City Hospital. A resident in medicine, William Bush, had a patient who required streptomycin for the treatment of tuberculosis, but he suffered from kidney failure as well. Bush knew that streptomycin was excreted almost entirely by the kidneys, and therefore might be lost before it could be effective in battling the tuberculosis. On the other hand, if the dose were increased to compensate for what was being excreted by the kidneys, it might be lethal. When Bush asked Kunin how much of a reduction in dose there should be, Kunin was embarrassed that he did not know.

This prompted Kunin in 1958 to make a study of the residence times of antibiotic doses being given to the many people with renal failure at the Peter Bent Brigham Hospital in Boston. A broad spectrum antibiotic such as chloramphenicol was administered intravenously, and then blood samples were taken periodically to determine the residual concentration. A coinvestigator tackled the mass of resultant data by computing the half-life (i.e., the time required for the body to reduce the concentration of the drug initially present by half). The first three papers summarizing these studies were rejected by reviewers for the *Journal of Clinical Investigation.* After what were probably fiery rebuttals by Maxwell Finland, the papers were accepted and all were cited subsequently more than one hundred times in the literature. In one paper by Kunin and Finland on dimethylchlortetracycline, the formula for determining the half-life is spelled out.[10] In radiation chemistry, half-lives had been established for many years, but their entry into medicine was delayed considerably.

PHYSIOLOGY DEVELOPMENTS

Parrot Fever

Doctors don't know everything, but sometimes they learn from unusual sources. An example is given in a 1979 issue of British *Medical Journal* where W. Evans told of a serious epidemic that had broken out in London's East End with fever, delirium, and pneumonia symptoms preceding death.[11] Sir Robert Hutchison was summoned to the bedside of an elderly patient, and after examining her he came downstairs to consult with the family's doctor. Neither of them knew what disease the patient had, and that was apparent to a little eight-year-old girl in the household. She tugged at Hutchison's coat and said, "I know what's wrong with Auntie; she's got what the parrot died of." The patient was taken to the hospital and examined by Dr. Samuel Bedson, the clinical pathologist, who ultimately was credited in 1930 with discovering and describing the bacterium of psittacosis, or what is commonly referred to as parrot fever.

A Factor in the Control of Epilepsy

George Kauffman, one of the most prolific technical writers of any age, tells a story about valproic acid, a dicarboxylic acid first mentioned in the literature in 1881.[12] It wasn't until 1963 that it was shown to have antiepileptic properties. In his 1961 thesis at the Universite de Lyons, Pierre Eynard described the synthesis of a series of derivatives of the compound called khellin. Later, at Grenoble, when he began a study of the pharmacodynamic properties of these compounds, their insolubility in water or solvents posed a real problem. Along with H. Meunier and Y. Meunier, he tried to dissolve the most active member of this series in valproic acid (for the Meuniers had previously used valproic acid as an intermediate in preparing a bismuth salt used in therapy). The acid dissolved the khellin derivative, and the resulting solution was submitted to a battery of tests. It showed a protective action against artificially induced epileptic attacks. At this point H. Meunier had occasion to use valproic acid again as a solvent for a different type of compound, and it also surprisingly protected rabbits against seizures. Eventually, it became clear that the common solvent valproic acid had been responsible for the antiepileptic effects!

Subsequently H. Meunier found that a coumarin derivative, which was difficult to dissolve, was soluble in valproic acid, and this solution also protected rabbits against seizures that were artificially induced. Meunier knew that this surprising result could not be attributed to coumarin and concluded that the antiepileptic properties of the solution were due to the valproic acid solvent.

ON THE VALUE OF UNRESTRICTED INQUIRY

There have already been many references in this book to the writings of Julius Comroe Jr. and readers who have a serious interest in serendipity should read all that he has written. In this section we will explore two of his articles on a number of medical advances.[13, 14] His theme was the familiar story of the man in ancient China who left his son in his straw hut and in the care of a small herd of pigs. Playing with fire, the boy ignited the hut and in the ensuing conflagration, nine little piglets were burned to death. But the succulent smell associated with the unfortunate deaths of these pigs prompted the man and boy to sample the meat. For thousands of years humans had eaten raw meat, but the boy and his father discovered that the roast pig was far preferable to the raw, thus serendipitously benefiting the human diet.

In the first of his articles Comroe described eight accidental situations that led to great medical discoveries, including the staining of tubercle bacilli, the Gram stain, the development of the stethoscope, and the properties of epinephrine and ephedrine. Two of his accounts of serendipitous developments share a common theme—unplanned visits to the laboratory during the night as described below.

The Full Potential of Sulfapyridine

Before penicillin became available, sulfapyridine was found to be effective against pneumonia. A lucky break in 1940 played a role in developing this compound's full potential. Lionel Whitby in England was testing the new drug on mice, and the regimen adopted was to treat the mice during the day only. However, Whitby had been to a dinner party one night, and before returning home dropped by the laboratory to see how the mice were getting along. He decided to administer another shot of sulfapyridine to them, and this particular batch of mice withstood the pneumonia better than the others had. It was not until a week later that Whitby made the connection with the additional application as the difference between success and failure.

"We'll Leave the Lights on for You"

About the same time, according to Comroe, A. V. Nalbandov was studying the effects on chickens of the removal of their pituitary gland. Within a few weeks after the operation, the birds would die, and this was in accord with the findings of two other investigators, A. S. Parkes and R. T. Hill. Nalbandov was about to drop the whole project when, for no apparent reason, ninety-eight percent of a group of birds survived for three weeks and many lived for as long as six months. But further discouragement ensued when other batches of chickens died after the removal of the pituitary gland. Then another group of chickens survived. A break in the mystery came late one night when Nalbandov was driving home along a road that took him past the laboratory. Even though it was 2:00 AM, the lights were burning brightly, so he stopped in to investigate. He assumed that a careless student must have left the lights on, so he turned them off. A few nights later he discovered that the lights had been left on again. Upon inquiring further into the matter, he found that a substitute janitor, whose job it was to close the windows and lock the doors, preferred to leave the lights on in the animal room so that he could locate the exit door. Further investigation showed that the periods in which the chickens survived had coincided with the times served by the substitute janitor. What followed was a series of controlled experiments that showed that the chickens without the pituitary gland survived very well so long as they were in a lighted room for two one-hour periods each night. It was learned that the birds in the dark did not eat and consequently developed fatal hypoglycemia.

Finding a Rat Poison

In the second of Comroe's "Roast Pig" articles he traces the origins of several discoveries, which probably would never have been made except for sheer accident. He tells, for example, how Curt Richter, professor of psychobiology at Johns Hopkins, discovered a rat poison. Richter was interested in the phenomenon of taste, particularly the ability of animals to select a proper diet by using their sense of taste. For a rat, this is most important because it is unable to vomit; therefore, it must be selective in what it eats. It happened that Herbert Fox, a research chemist at DuPont, noticed one day that as phenylthiourea was blown through his laboratory in the form of a fine dust, some workers complained of a bitter taste in their mouths while others didn't notice it. Fox was interested enough in the findings to determine what proportion of people might be affected by phenylthiourea. Only fifteen percent found it extremely bitter. When Richter heard of Fox's study, he decided to determine how rats would respond to phenylthiourea. He tried its effects on six rats; the dose was minimal, being only what would stick to the end of a toothpick. He observed the rats for a half hour to see whether there would be any signs of distaste, such as using their paws to clean off their tongues. Nothing happened and Richter stopped his observations, but the next morning he discovered that all the rats had died. Their chests and lungs had filled with fluid. Further studies showed that only one to two milligrams of phenylthiourea were needed to kill a rat.

The Discovery of Dramamine

Comroe tells of another noteworthy event that had totally unforeseen results. In the 1940s L. N. Gay and P. E. Carliner, allergists at Johns Hopkins, were testing the effects of antihistamines on their patients. The Searle Company sent them a sample of dramamine, a prospective antihistamine that was being tested along with other drugs. One of the patients noticed that since she had been taking dramamine, she had no symptoms of the car sickness that occurred with every (even short) trip she took. Dr. Gay tested these observations by switching her medication alternately from dramamine to a placebo and confirmed that dramamine was controlling her car sickness. A large-scale study of the effectiveness of dramamine came about because of the military's experience with seasickness during World War II. The D-Day landing in Normandy several years earlier demonstrated the need for seasickness control and the Army was happy to cooperate in "Operation Seasick," supervised by Dr. Gay with one thousand five hundred soldiers on a troopship for twelve days in the rough North Atlantic. This study showed that dramamine could be effective in both preventing and relieving motion sickness.[15]

THE UNPREDICTABILITY OF BREAKTHROUGHS

A sequence of policy statements led to a convincing demonstration of the necessity of basic research in medicine. First, President Lyndon Johnson made a statement in 1966 that we must "make sure that no lifesaving discovery is locked up in the laboratory." This was a thinly veiled attack on basic research, the inference being that the nation could not waste money on impractical research. One consequence of this was that the Defense Department published "Project Hindsight," a retrospective analysis of how twenty important military weapons were developed. Its contention was that scientists were most effective when their studies were confined to the goals of the sponsor. Defense-sponsored university research was shown to have played only a minor role in the development of these twenty military weapons. To counter these attitudes, various medical researchers and other scientists prepared reports that showed the importance of basic research.

Two men were highly critical of the players on each side of the controversy, because each was basing its position on anecdotal occurrences. These men were Julius Comroe and Robert D. Dripps, professor of anesthesia and vice president for health affairs at the University of Pennsylvania in Philadelphia. They began by polling their colleagues, asking them to rank the most important medical advances since 1940. Topping the list was cardiac surgery, including open-heart repair of congenital defects and replacement of diseased valves. Comroe and Dripps then made a massive study of approximately four thousand technical articles going back over one hundred years or more that played a necessary role in making open-heart surgery possible. Included were X rays, studies of anesthesia, blood typing, blood clotting, catheterization, and a host of other subjects that would not readily come to mind as necessary factors in heart operations. In short, the most important finding was that forty-one percent of all the work judged to be essential or crucial for later clinical advances was not clinically oriented at the time this research was done. Their study is a convincing and illuminating work on the unpredictability and necessity of basic research.[16]

REFERENCES

1. Mellanby, John. 1956. Sir Edward Mellanby (1884–1955): The man, research worker, and statesman. *Ann. Rev. of Biochem.* 29: 1–28.

2. Altman, Lawrence. July 9, 1991. *New York Times.*

3. Comroe, J. H. 1976. Tell it like it was. *Amer. Rev. Resp. Diseases.* 113: 667–676.

4. Barnett, H. J. M. et al. *Stroke: Pathophysiology, Diagnosis, and Management.* New York: Churchill Livingston Publ. 1992.

5. Lindenmann, J. 1989. *Current Contents (LS).* Nov. 13, pg 19, referring to Isaacs, A. and Lindenmann, J. 1957. Virus interference. 1. The interferon. *Proc. Roy. Soc. B.* 147: 258–273.

6. Cleland, W. W. 1985. *Current Contents, Feb. 4,* pg 18, Feb. 4, referring to Cleland, W.W. 1967. The statistical analysis of enzyme kinetic data. *Advan. Enzymol. Relat. Areas Mol.* 29: 1–32.

7. Dale, Henry H. 1963. Pharmacology during the past sixty years. *Ann. Rev. of Pharmacol.* 3: 1–8.

8. 116th Edition of Statistics Abstracts of U.S., 1996.

9. Kunin, Calvin. 1985. *Current Contents (CP).* Sept.16, pg 20, referring to Kunin, Calvin M. 1967. A guide to use of antibiotics in patients with renal disease: a table of recommended doses and factors governing serum levels. *Ann. Intern. Med.* 67: 151–158.

10. Kunin, Calvin and Finland, M. 1958. Demethylchlortetracycline. A new tetracycline antibiotic that yields greater and more sustained antibacterial activity. *New Engl. J. of Med.* 259: 999–1005.

11. Evans, W. July/Dec. 1979. An incorrect reference was made to *Brit. Med. J.* Actual location unknown.

12. Kauffman, George. 1982. *Education in Chemistry.* 19:168.

13. Comroe, Julius Jr. 1977. Roast pig and scientific discovery. Part 1. *Amer. Rev. of Resp. Diseases.* 115: 853–60.

14. Comroe, Julius Jr. 1977. Roast pig and scientific discovery. Part II. *Amer. Rev. of Resp. Diseases.* 115: 1035–1044.

15. *University of Johns Hopkins Medical Bulletin.* 1949: 84–88.

16. Comroe, Julius Jr. and Drips, R. D. 1976. Scientific basis for the support of biomedical science. *Science.* 192: 105–111.

Chapter 7

Soil Microbes, Anesthetics, Sunken Submarines, and Other Biological Tools

Common ground among scientific disciplines is always expanding; therefore the distinctions are less clear than they were years ago. In writing this book, a recurring problem was to decide the proper location for individual facts because they might fit well in any of several locations. This chapter has biology as its principal theme but much of it might well have been included in chemistry or other fields.

There's Fungus among Us

Credit for the most complete response to my letter addressed to recently elected members of the Royal Academy and to the National Academy of Sciences must go to T. Kent Kirk, director of the Institute for Microbial and Biochemical Technology at the USDA Forest Products Laboratory in Madison, Wisconsin. His letter was filled with interesting narratives, great examples of serendipity, and real insights into the genesis of discoveries. Kirk's career was shaped largely by his early association with a professor of forest pathology, Otto Wasmer, at what was then Louisiana Polytechnic Institute (Ruston, Louisiana). Kirk was only nineteen when he came under Wasmer's influence and had no idea of becoming a scientist, but he was "fascinated by the intricacies of the interactions between microorganisms and plants; at the time I was especially intrigued with how some fungi form a synergistic relationship with the roots of trees (mycorrhizae)." He even submitted a term paper on mycorrhizae in an essay contest, won the first prize of fifty dollars, and saw his first publication at the age of twenty-one. Perhaps even more remarkable was that when his schedule did not permit him to sign up for any microbiology courses, Wasmer gave him private lessons over coffee. Kirk then did graduate work in plant pathology with Arthur Kelman and Ellis Cowling at North Carolina State, both of whom insisted that their students have good chemistry backgrounds. As a consequence Kirk graduated with a double major with biochemistry and plant pathology at the PhD level. At the urging of Cowling, he took a postdoctoral fellowship in Sweden as Cowling had done years before.

Kirk spent one and a half years studying lignin chemistry in the excellent laboratory of Professor Erich Adler at Chalmers Technical University at Gotenborg, after which he received a position at the U.S. Forest Products Laboratory where he has kept himself and other workers busy for over twenty-five years.

T. Kent Kirk

Regarding serendipity, here is Kirk's account:

> Perhaps the most singular example of serendipity in our work was in the discovery that lignin-degrading fungi must be in a "secondary metabolic" or nutritionally stressed state before they will degrade lignin. In the early 1970s I had decided to synthesize radioactively labeled lignin to develop an unequivocal assay for lignin biodegradation. It was expensive in time and materials, and I was dismayed that the fungi ignored our precious lignin when it was added to growing cultures in defined liquid media. We eventually got the cultures to degrade the lignin, albeit slowly, by adding the lignin to solid wood and allowing the fungi to decay the wood; this was a great relief at the time because it showed that our lignin at least was being recognized by the degradative enzyme system of the fungi. But why was it not recognized in defined liquid media? We were trying various approaches and imagining all kinds of answers when the problem solved itself; in one experiment, for no reason that was immediately apparent, we saw the first efficient degradation of the labeled lignin in a chemically defined medium. We were then able to deduce from the lab notes that the culture medium had been made up ten times more dilute than it should have been (an easy mistake to make), and subsequent experiments showed that by limiting nutrient nitrogen alone we could trigger the ligninolytic system. I suppose we would have eventually made that discovery, but serendipity saved us precious weeks. Later work showed that carbon-limitation, and under some conditions sulfur-limitation, also triggers the ligninolytic system. (This all seems to make sense intuitively, because wood is not a rich source of nutrients, but an intellectually satisfying explanation for the connection between lignin degradation and secondary metabolism has not been advanced.)

Kirk described a second serendipitous occurrence where a fungus he and Terry L. Highley were studying had been isolated from a decaying wood chip in Maine (*Peniaphora "G"*). Their interest was still in lignin degradation, and this fungus was the fastest and most aggressive lignin-degrading fungus they had screened. Furthermore, its copious production of asexual spores made it easy to handle in the laboratory. It was interesting also because of its high optimum growth temperature (40° C) and it produced no laccase, both aspects being rare. Meanwhile Harold Burdsall, a mycologist at the Forest Products Laboratory, had determined that this fungus was a new species (*Phanerochaete chrysosporium Burds*). As a footnote, back in the 1960s Karl-Erik Eriksson in Stockholm had chosen another wood-decay fungus (*Chrysosporium lignorum*) because of its fast growth rate and was studying cellulose degradation with it. By the late 1970s mycologists in Sweden had renamed it *Sporotrichum pulverulentum*. Subsequently, Burdsall's studies revealed that *S. pulvurulentum* and *P. chrysosporium* are the same species. The result was that more is now known about both cellulose and lignin degradation in this species than would have been the case if each study had been based on different organisms.

A third case of serendipity related by Kirk developed from the use of *P. chrysosporium.* A friend and colleague of Kirk since postdoctoral days, Knut Lundquist of Chalmers, was studying ligninolytic cultures of this organism in an attempt to identify specific degradative reactions that might serve as enzyme assays. Low-molecular-weight model compounds were added to cultures of the fungus, and after certain periods of time samples were analyzed. The first examination was by thin-layer chromatography with spray reagents that developed colored spots on the plates. One day Lundquist, who was an excellent lignin chemist, came to Kirk in a high degree of frustration because all of the cultures yielded a rosy red spot with a certain reagent. He was convinced from past experience that the spot was veratryl alcohol. A quick check showed him to be correct, but how could veratryl alcohol have been produced as a metabolite from so many diverse compounds as had been used in this study? Their immediate suspicion was that a technician had inadvertently put veratryl alcohol in all of the cultures, but that was not the case. Further study showed that *P. chrysosporium* produces veratryl alcohol as a secondary metabolite; later it was learned that veratryl alcohol is produced by most (if not all) lignin-degrading basidiomycetes. Furthermore, it protects the key degradative enzyme, lignin peroxidase, from destruction by H_2O_2. Also it apparently induces the whole ligninolytic enzyme system. This important cluster of serendipitous findings had a profound effect on the current understanding of nature's way of degrading wood.

A Nobel Prize Beginning with a Need for an Anesthetist

Winning the Nobel Prize was the culmination of decades of brilliant research by Jens Skou that depended in part on elements of chance. His letter to me concluded with the observation, "...had it not been for chance I would probably today have been a surgeon."

For the benefit of those who are not scientists, there is a convention used to describe the action of enzymes. The suffix "-ase" is added to the term describing the enzyme's activity: for example, when an enzyme catalyzes the splitting up of a compound known as an *ester,* it is described as an *esterase.* Similarly, an enzyme catalyzing the splitting of the molecule *adenosine triphosphate,* commonly referred to as ATP, is called an *ATPase.* There is probably not a more important enzyme in the body than ATPase and as the following narrative shows, an understanding of the role of two ions in membrane channels was certainly not predicted. The two ions are derived from the metals *potassium* (K^+) and *sodium* (Na^+).

Three researchers shared the 1997 Nobel Prize in chemistry. Paul Boyer (University of California, Los Angeles) and John E. Walker (Medical Research Council Laboratory of Molecular Biology, Cambridge, UK) shared half of the award for elucidating the enzyme mechanism by which ATP synthase (ATPase) catalyzes the synthesis of adenosine triphosphate (ATP), sometimes referred to as the energy currency of living cells. Skou (Aarhus University, Aarhus, Denmark) received the other half of the award for his discovery of the first molecular pump, an ion-transporting enzyme called Na^+,K^{+-} ATPase.

What led Skou into this remarkable work was his interest in surgery in the late 1940s. There was no anesthetist in the ward where he worked so to avoid the use of ether, they used spinal and local anesthesia whenever possible. Skou developed an interest in the mechanism of action of local anesthetics and decided to pursue the subject as a thesis topic before continuing his surgical career.

Skou was aware of previous work that seemed to relate to the action of anesthetics, but first some background in "monolayers" is essential here. If a drop of oil is placed on water, it spreads out (not dissolves). The film that is formed has a structure and if a barrier is set up to compress it, the structure of the oil film changes, including whatever compound is dissolved in it. Irving Langmuir pioneered in the study of compressed surface films, and he showed that the area of an oil-soluble component of a surface film could be calculated from the pressure buildup in that film. Research by J. H. Schulman had shown that certain drugs in an aqueous phase that was covered by an oily film could penetrate into the oily film. Also it had been shown that the activity of a drug could be influenced by changes in pH, and there was a correlation between the pH and the penetration of the drug into the oil monolayer. With these considerations in mind, for a drug (e.g., an anesthetic) to be active it seemed necessary for it to penetrate the oily layer of a cell membrane.

We pick up the story with Skou's *Citation Classic:*[1]

> In the beginning of the 1950s I was interested in the effect of local anesthetics and had found that the increase in surface pressure by penetration of local anesthetics into a monolayer of lipids extracted from nerves correlated to their local anesthetic effect. This raised the question: can an increase in surface pressure from penetration of a local anesthetic into the lipid part of a nerve membrane influence the configuration of proteins in the membrane and thereby block the trigger mechanism in the Na^+ channels?

He needed a monolayer of a lipid protein with enzyme activity to test the effect of penetration of a local anesthetic. Circumstances led him to study the microsomes of crab nerves, but the activity varied from one preparation to another for no apparent reason. Finally, after three months of work he found that K^+ in the test solution increased intensity, at which point he went on a summer holiday to forget the whole thing. When he returned, he repeated the experiment but it failed. Ultimately he learned that both Na^+ and K^+ were needed with the ATP experiments, each ion being pumped across the cell membrane.

Skou realized later that his choice of crab nerves was highly fortuitous, because it is one of the few tissues where the Na^+, K^+-ATPase activity is revealed without the use of detergents. If he had used a mammalian tissue there would probably have been a major problem; most of the enzymic activity is hidden because of vesicle formation of the plasma membrane fragments, and the vesicles must be opened, probably with a detergent, in order to see the combined effect of Na^+ and K^+. With crab nerves, however, there was no such need for a detergent.

A Chemistry/Physics/Biology Interface

Knowledge of the interrelations between Ca^{++} and the proteins in blood serum came about in a unique way. Baird Hastings happened to be sitting next to Jared Morse, a friend in the Department of Physics at the University of Chicago, during a lunch break in 1928; Hastings was discussing the problem of blood serum and calcium. In a continuing controversy over the matter three factions held disparate views, and on this particular day Hastings was morose because there seemed to be no resolution to the problem. Hastings was among those who held that there was an equilibrium between the blood serum and the calcium concentration in the bones. Another group maintained that the bones were supersaturated with regard to calcium. The third group maintained that there was an undersaturation. No one knew what the solid phase in the bone was. Morse asked what was known about the structure; Hastings replied that it wasn't crystalline. Morse commented that everything has some sort of structure, and he suggested that Hastings give him a sample of bone for X-ray analysis. The study that developed required a considerable time, but it was definitive in that it showed the bone salt was not a mechanical mixture of $CaCO_3$ and $Ca_3(PO_4)_2$ and it was not $CaHPO_4$; it was a carbonate apatite.

With the question of the solid phase resolved, next was the Ca^{++} concentration in plasma to be defined, and serendipity figured in the solution. It was 1932 when Hastings lured Franklin McLean into his Department of Medicine laboratory to set up a study of the effects of various ions on the performance of a rabbit's heart. Work had to be done in a laboratory maintained at the warm temperature of 38° C. When the researchers were transferred to the Department of Physiology, no room there was maintained at that temperature so their experiments with the rabbit heart were discontinued. But McLean recalled that twenty years earlier it had been shown that the frog heart was sensitive to small changes in the Ca^{++} concentration of fluids being passed through the heart. And since there was no requirement that a frog's heart be kept at an elevated temperature, they used it as the test subject. They were able to determine the Ca^{++} concentration of the blood serum under various conditions and eventually demonstrated a relationship between Ca^{++} and proteins. The result was a monogram from which one could read off the Ca^{++} provided the total calcium and the protein concentration of serum were known. In summary, a controversy concerning three competing views about calcium and blood serum was resolved largely through the expertise of a physicist, and despite the lack of a room maintained at a high temperature.

The Unusual Origin of the Pap Smear

"Teratology" is the term used to denote abnormal embryonic development, and it has become well-known in recent years because of the unfortunate deformities of infants caused by thalidomide. Much remains to be learned about embryonic development, but it has been known that deformities can be caused by the interruption of certain processes during development.

An investigator in the nineteenth century wanted to test the effect of all the metals in the first column of the periodic table on deformities of marine organisms. Sodium and potassium are in that column, and both elements occur abundantly in seawater. Lithium is also in the same group and therefore was used in studies of development. In 1906 Charles R. Stockard began his work on teratology by studying the effect of lithium chloride solutions on the development of eggs of the killifish, *Fundulus heteroclitus.* Stockard detected abnormalities including slow development, defective eye formation, shortening of the body, and forking of the tail, but he could not be certain that the effects on the killifish eggs were specific for lithium. The next year he used magnesium solutions with the eggs and published an article describing the production of cyclopean eyes. He suggested that the effect might be strictly chemical and not resulting from a change in osmotic pressure, which had been known to affect development. Stockard also pointed out the similarity of these cyclopean eyes with the anomalies known to occur occasionally in humans. In 1909 he questioned whether the malformation caused by magnesium ions might have resulted from an anesthetic effect attributed to magnesium. And so it was that he studied the effect of various known anesthetics, among which was alcohol, that caused deformities in the central nervous system, the eyes, and the ears in a large percentage of the specimens tested. Alcohol had a more pronounced effect than ether or chloroform. Subsequent studies with alcohol and its effects on guinea pigs led to his universal conclusion:

> The primary action of all the treatments is to inhibit the rate of development, and the type of deformity that results depends simply upon the developmental moment at which the interruption occurs.

During that time Stockard collaborated with George N. Papanicolau. One of the projects was to improve their control conditions by placing male and female guinea pigs together only when the latter were in a particular phase of estrus, to avoid preferential matings. To standardize the staging of the estrus cycle they studied vaginal smears, and because Papanicolau had considerable experience in cell cytology, he recognized the presence of cancer cells occasionally. He then proposed that these vaginal smears be used as an early indicator of cervical cancer and thus the Pap smear became known as the basis for the clinical detection of cancer. One could hardly imagine a train of events beginning with a study of ions as a causative agent for birth deformities leading to a method for the detection of cancer. That is how science seems to work despite all logical efforts to make it reasonable.

Oops! It's a Male!

In the story of Geoffrey Burnstock can be found several themes that recur throughout this book. The first is that he was considered an outsider to the field in which he made his major contribution; approximately twenty years elapsed before his finding was accepted generally. The second is that a chance occurrence was responsible for the finding. A complete account is to be found in an article by M. J. Rand and F. Mitchelson.[2]

Burnstock, head of the Department of Anatomy and Developmental Biology at University College, London, was not primarily a pharmacologist and therefore less likely to be stuck in the ruts of traditionalism governing research on nerve responses. He developed the "purinergic nerve hypothesis" which was an addition to the existing adrenergic (responsive to adrenaline) and cholinergic (responsive to cholinesterase) hypotheses. Studies on intestinal smooth muscle have traditionally formed the basis for developments in pharmacology and physiology concerning the autonomic nervous system. Other organs containing smooth muscle can also be used. A procedure common to all such studies is that specific compounds can be used to block the response of a nerve so that an adrenergic response can be distinguished from one that is cholinergic. With that as the background, here is the story.

It begins with the vas deferens, an organ that contains smooth muscle, and how its use led to Burnstock's discovery. Seid Hukovic, a Yugoslavian, was working in Oxford in 1959 with the late Professor J. H. Burn. Burn had been working hard to find evidence for a cholinergic link in adrenergic transmission and felt that it might be wise to repeat and extend some observations made by a previous Yugoslavian visitor to the laboratory. The earlier study involved the effects of an alkaloid, physostigmine, on the response of the rabbit uterus to hypogastric nerve stimulation. Hukovic intended to repeat that work, but when he went to the animal house there were no female rabbits in stock. There were plenty of guinea pigs, however, and he made the decision to substitute a female guinea pig for a female rabbit for the study at hand. But once he had slaughtered the animal, he used the vas deferens of the animal with its hypogastric nerve attached, reasoning that a vas deferens was roughly analogous to a uterine horn.

If a nerve supplying a tissue containing smooth muscle is stimulated, it releases a transmitter which mediates the response. Classically, the transmitter released would be either acetylcholine or noradrenaline, depending on the type of nerve. With the vas deferens as a source of smooth muscle, however, the response to excitation in the presence of various specific blockers was abnormal. Subsequently Burnstock derived the hypothesis that the nerves in the vas deferens were responsive to a purine (probably ATP); therefore he used the term "purinergic' to describe them. A problem with the study of ATP as a transmitter was the lack of a potent and selective antagonist. Also a bias against the whole study was that it was to a large extent the work of Burnstock, a zoologist, although he had published papers in pharmacology. Over the course of time, Burnstock's hypothesis has become accepted by even the pharmacologists. The origins of the discovery were unusual—no female rabbits available, then a mistake concerning the sex of a guinea pig being used as a substitute—but truth prevailed.

Never Trust an Expert

Research on genetic manipulation of plants by the use of plant protoplasts has been accelerated, thanks to a chance occurrence in the career of Edward C. Cocking. The acceleration would have been faster if he hadn't heeded the advice of an expert.

Cocking was head of the Department of Botany and the Plant Genetic Manipulation Group at the University of Nottingham (UK). In 1956 while he was studying the nitrogen metabolism of plants, government policy forced him to perform research on the development of vaccines against plague. This brought him into contact with workers pioneering the use of bacterial protoplasts, and he was subsequently involved with enzymatic degradations of cell walls. Cocking knew that there was a wide range of commercially available cellulases (enzymes that degrade cellulose, a constituent of plant cell walls), and he asked D. B. Whitaker of the Canadian National Research Laboratories, Ottawa, for a sample of a cellulase he had prepared from the fungus *Myrothecium verrucaria.* Whitaker graciously supplied the enzyme but sent along a covering letter that sounded a discouraging note, suggesting that the degradation sought by Cocking would be slow because the cellulose in cell walls tends to be crystalline. Accordingly, Cocking placed Whitaker's preparation at the bottom of the heap and proceeded to test all the other cellulases he had accumulated. None worked. Finally, he tried Whitaker's enzyme and found it released the protoplasts Cocking had been looking for. It was the only active preparation he had available for several years.

Having read numerous accounts detailing instances in which experts have been wrong in their predictions, I have concluded that our judgments might be too simplistic. A definition of an expert according to *Webster's New Collegiate Dictionary,* 1979, is "One who has acquired special skill in or knowledge of a particular subject." To be complete, that definition should include a caveat *"at a point in time."* For example, Einstein made a statement in 1932 to the effect that there would be no such thing as atomic power, even though he was the developer of the equation $E = mc^2$. Had he known that the process of fission was to be demonstrated shortly thereafter, he would not have made that statement. It is well to remember that a man might really be an *expert,* but not a *prophet.*

Discovery through Delay

Waiting for someone else to finish using a centrifuge is not a part of a normal research plan, but it solved a problem for Arne Boyum (Norwegian Defense Research Establishment, Kjeller, Norway). His article describing the event was cited as a reference almost six thousand times in fourteen years.[3] In his early attempts to separate mononuclear cells and granulocytes from human blood, Boyum constructed a large centrifuge which did not perform well. His approach had been to establish a density gradient with the use of highly polymeric compounds, but with this scheme density and viscosity could not be varied independently. He then tried two materials—an X-ray contrast medium to adjust the density and sugar polymers to control viscosity. It was then that the accidental delay totally changed the situation.

The density gradient was already loaded, but someone else was using the centrifuge, and Boyum had to wait his turn. During the wait, he noticed that the red cells started to aggregate at the interface and then fell rapidly to the bottom. This prompted him to investigate further this situation with all of the variables. Ultimately his careful research and patient observations led to the development of a suitable centrifuge technique. In all, the process took three and a half years, but enabled Boyum to obtain a pure suspension of mononuclear blood cells.

New Year's Card Leads to Discovery of a New Plant Species

Dr. Hugh Iltis, professor emeritus of botany at the University of Wisconsin, in early 1976 sent a New Year's card (really a poster) to botanists around the world. It was his drawing of *Zea perennis,* a perennial grass thought to be extinct, which he so marked on the drawing. Two years later he received a handwritten note from Anthony Pizzati who worked for a messenger service in New York City. Pizzati said a Mexican friend had found "*perennis*—the long-lost original corn." Would he be interested? So Iltis dispatched graduate student John Doebley to the task and suggested that he get some of the seeds to find out what Pizzati was really talking about.

What had happened was that Iltis's poster had been placed on a bulletin board at the University of Guadalajara by a professor who, slightly irked, had urged her students to go find this *teosinte* and prove that the gringo Iltis was wrong. Graduate student Rafael Guzman went to the original locality in the mountains of Jalisco and by the second day had dug up a sterile plant; in the next two months it grew into the long lost perennis. Guzman went even further. Within another month he had learned from a fellow student that *Zea perennis* was growing in another location; when he located that population he sent the seeds to Iltis. When grown in Wisconsin this *teosinte* turned out to be not only a perennial but one with half the number of chromosomes. It was a totally new species which was called *Zea diploperennis.* Unlike *Zea perennis,* this species freely interbreeds with corn which raises the possibility that the crop could be grown for several years from one rootstock. If corn could be grown as a perennial, like apples or hay, it would lead to a tremendous savings in soil erosion. In addition, *Zea diploperennis* is immune to, or tolerant of, seven corn viruses and is the only member of *Zea* that is immune to three of them.

Another consequence of the poster was the establishment in 1987 of the three hundred fifty thousand acre Sierra de Manantlan Biosphere Reserve. This is the first reserve established principally for the preservation of a wild crop relative. It is only one percent the size of Wisconsin but it has one-third again as many higher plant species—2,650. Also, the poster led to the establishment of close ties between the University of Guadalajara and the University of Wisconsin. More than a dozen students have gone to Wisconsin for graduate study, including Rafael Guzman.

Bioluminescence

Of all of nature's wonders, bioluminescence might be the most hypnotic. Certain microorganisms emit light spontaneously, most often as the result of physical agitation, and when large concentrations of these organisms light up, they make a spectacular sight. Anyone who has seen intense bioluminescence, either at sea on a dark night or in a laboratory vial, will remember the experience. It is no wonder that Beatrice (Bea) Sweeney made it her life's work once she began her study of *Gonyaulux polyedra.* In an earlier chapter it was mentioned that she began a study of bioluminescence because she did not have the specialized equipment she needed when she went to her new post. In her experiments with bioluminescence that followed, she benefited from serendipity.

In studying the effect of various light/dark regimens on the luminescence of *Gonyaulux,* Sweeney would illuminate a culture for a known period and then place aliquots of it in test tubes to be kept in the dark. Periodically, she would remove a test tube from the dark, agitate it in a standard fashion, and measure its luminescence. The longer the dark period, the more light was produced, but only up to six hours, after which there was a decline in light production. Sweeney still had cultures in the dark and did not want to discard them even though her initial goal seemed to have been achieved. So she continued with the hourly studies and found that after twenty-four hours of darkness, the light production of her cultures was almost as intense as it had been initially. This circadian effect was new to her, so she presented her results at a technical meeting.

Woodland (Woody) Hastings was at that meeting. He was alert to the circadian effect she had found because Frank Brown, a confrere at Northeastern University, was experimenting with the rhythms of crabs and potatoes. Sweeney and Hastings ultimately became close collaborators in research on bioluminescence.

Sweeney became aware of a specific light effect on bioluminescence in an unusual way. Much of her data had been amassed at Scripps Institution of Oceanography, but when parallel experiments were run at Caltech, the results were not in total agreement. Subsequent investigations showed that the lights in the temperature-controlled rooms at Caltech were run at an overvoltage, and that produced more blue light than that used at Scripps.

Bioluminescence was involved in an unintended fashion at the Naval Research Laboratory (Washington, DC) in the 1980s. A postdoctoral fellow in the author's laboratory, Arthur Stiffey, was studying the luminescence of *Pyrocystis lunula* and had devised a two-hour bioassay based on its performance. At that time I was studying the literature concerning trichothecenes, the class of compounds known in the popular press as "Yellow Rain." This study was prompted by the charges made by the U.S. Department of State that Yellow Rain compounds were being used as chemical weapons against South Vietnam, Laos, and Cambodia. Clearly there was a need for a rapid method to detect the presence of these toxic compounds in order to prevent adverse health effects to American soldiers fighting there. When I discussed the subject with Stiffey, he raised the question, "What would happen if you added a Yellow Rain compound to a *Pyrocystis* culture?" My reaction was that there seemed to be no reason for introducing a terrestrial fungal metabolite into a marine dinoflagellate culture, and I asked, "Why would you do that?" Art's response was simply, "Because it would be easy to do" (much in keeping with the philosophy of "doing the quick and dirty"). To our surprise, the "Yellow Rain" compounds caused a significant decrease in bioluminescence. Subsequent research showed that with a simple leaching procedure, requiring no workup, 0.25 microgram of T-2 toxin could be detected on vegetation in this two-hour assay. One knock on the method was that it would not identify the compound. Another effort to detect these compounds at NRL was the gas exchange approach, but the Defense Department chose to rely solely on the advanced chemical technique, mass spectrometry coupled with gas chromatography.

Art Stiffey

Submarine Sandwiches and Microbial Cultures

Stories with happy endings always have an appeal, and this one began on a sour note and ended with a scientific breakthrough.

The Woods Hole Oceanographic Institution (Falmouth, Massachusetts) has a minisubmarine, *Alvin,* which has been a great asset in many oceanographic studies. On October 16, 1968, it was setting on the deck of its mother ship when a restraining cable broke and the sub plunged into the sea. With the hatches open (fortunately no one was aboard) it filled with water quickly and sank to the bottom of the Atlantic Ocean, 1,540 meters deep. Its position was 135 miles southwest of Woods Hole. In the submarine were the lunches of the crew, including two thermos bottles of bouillon and a plastic box containing sandwiches and apples. No one could have suspected at the time of the sinking that these crewmen's lunch items would play such an important role ultimately.

A salvage operation ten months later recovered *Alvin* with all of its trapped contents. The general appearance of the food items indicated that they were well preserved and the microbial studies made of them confirmed that. They had been subjected to a pressure of 150 atmospheres and a constant temperature of 3° C; there was no evidence of a reducing condition nor a lack of dissolved oxygen. The sandwiches, wrapped in wax paper, appeared to be in excellent condition considering their long exposure to salt water. When pieces of the bread were streaked on nutrient agar, bacteria and molds grew profusely. When submerged in sterile seawater, the meat spoiled with a putrefactive smell within four weeks at 0° C. In short, the food items were well preserved in the high pressure of the deep sea and whatever microorganisms were resident in them remained dormant under those conditions.

There were several consequences of this unique occurrence:

It cast a huge shadow over the increasing possibility of the ocean being used as a dumping ground for urban sewage and wastes (there was no Environmental Protection Agency in existence at that time). Proponents of ocean dumping felt that the organisms in the sea would effectively destroy these wastes with few adverse consequences. The experience derived from the *Alvin* saga proved that was not the case, and ocean dumping was obviously not the answer.

A whole new emphasis on hyperbaric research developed, leading to significant health benefits. There had been basic laboratory studies of the effects of high pressure on microorganisms, but when attempts were made to conduct such studies in the sea, they were oriented toward capturing deep sea samples and bringing them to the surface without changing the pressure. These were difficult tasks, but what was learned was to duplicate what happened with *Alvin* (i.e., prepare the necessary cultures in the laboratory, deploy them in containers able to withstand the ambient pressures, and ultimately retrieve them). Then a comparison could be made of the submerged cultures versus those kept at the same temperature but at atmospheric pressure. Details of the *Alvin* sinking and the results obtained can be found in an article by Jannasch.[4]

One of the notable health developments resulting from this incident was the use of hyperbaric chambers for the treatment of gangrene. Pressure had been shown to inhibit certain microorgansims, such as those found in gangrene.

Would it not be reasonable to impose high pressures to body members, particularly the legs and feet, and thereby destroy the gangrenous organisms in them? The very beneficial result has been that many amputations have been avoided for diabetic patients having severe circulatory problems.

Before leaving the subject of *Alvin,* it is worth noting that the name is derived from its designer, the late Allan Vine. The mother ship for the little submarine was named *Lulu* after Vine's mother. Allan Vine was a pleasant, unassuming, and interesting man. I recall an incident when I was to pick him up at National Airport one morning in the late 1960s. All of the passengers seemed to have left the plane but there was no sign of Vine, and I was beginning to think that he had missed the flight. Eventually he showed up in the company of the pilot, the reason being that Vine was interested in learning the angle of glide path and the speed of the plane on landing so that he might be able to calculate the force involved in the plane contacting the ground. He wanted to compare it with the force of the minisub banging against a ship if it were hit by a large wave.

Unexpected Help from a Physician

Fritz Went is best known for his research on the effect of light intensity and the length of day on the growth of tomato plants. How those studies came about was highly unusual and certainly unplanned. Went was at the California Institute of Technology and was frustrated because of the lack of incubators there. Students and faculty members were competing for times to use the incubators, and so it was that when Went paid a visit to his retired physician he was bemoaning his fate. The physician, H. O. Eversole, happened to be an orchid fancier and offered to make an incubator suitable for Went's use. At that time in the early '20s, incubators customarily operated at a fixed temperature and that was satisfactory for Went's contemporaries who were preoccupied with biochemistry. Eversole had reasoned that the proper culturing of orchids might depend on alterations of both the temperature and light intensity. After all, when the sun goes down so does the temperature. His own incubator provided the options required to mimic nature's circadian rhythm, as did the one he made for Fritz Went. Having the luxury of an incubator for his own use was great, but having one of the type made by Eversole was even better. With it he studied the length of day required for the optimum growth of tomato plants and was able to explain why tomatoes did not do well in northern climates.

Intelligent researchers seem to make progress despite obstacles in their path. An example of this was evident when Went participated in the expedition of the *Alpha Helix* in 1967, and his equipment was impounded by Brazilian customs officials. So he spent his time in the Amazonian rain forest, striving to understand what was going on in the upper soil layer where a litter layer of dead leaves and branches was completely pervaded by tree roots and fungal hyphae, but no mushrooms. He found that there was a tripartite system in which the fungi digest the litter and pass much of the extracted nutrients back to the tree roots, closing a nutrient cycle without which a rain forest could never exist on the very poor, leached soils of most of the Amazonian basin.

Went must have been a fascinating person, judging by his reminiscences in the *1974 Annual Review of Plant Physiology.* Among his views was this gem:

> I am very little impressed by complicated and clever theoretical or mathematical constructions; in fact, I don't understand many of them. Nor can I follow or accept statistical analysis; if the facts don't speak clearly for themselves, no statistical treatment will make them palatable.

He was involved with some of the early work on auxins and told of controversies he had with others at that time:

> I well remember endless discussion on why the growth-promoting principle of the oat coleoptile was not a spirit and weightless, ghostly and immaterial. It induced me to measure the diffusion content of this "spirit," which indicated that auxin had a molecular weight ranging between three hundred and four hundred, which made it impossible to classify as a "spirit."

REFERENCES

1. Skou, Jens. 1981. *Current Contents (LS).* May 18, pg 20, referring to Skou, J.C. 1965. Enzymatic basis for active transport of Na^+ and K^+ across cell membranes. *Physiol. Rev.* 45: 596–617.

2. Parnham, M. J. and Bruinvels. J., eds., 1986. *Pharmacological Methods, Receptors, and Chemotherapy.* Amsterdam; New York: Elsevier Publ. Rand, M.J. and F. Mitchelson. The guts of the matter: Contribution of studies on smooth muscle to discoveries in pharmacology. Vol. 3.

3. Boyum, Arne. 1982. *Current Contents (LS).* Nov. 6, pg 20, referring to Boyum, A. 1968. Isolation of mononuclear cells and granulocytes from human blood. *Scand. J. Clin. Lab. Invest.* 21 (Suppl. 97): 77–89.

4. Jannasch, Holger. 1971. *Science.* 171: 672–675.

CHAPTER 8

LEARNING FROM ANIMALS AND INSECTS

It is normal practice to use animals, insects, and microorganisms as subjects of study in toxicity determinations. But they also have unique capabilities that are being recognized by scientists. This chapter contains examples of that and instances of discoveries stemming from unusual circumstances.

Several Observations Late in Coming

Animals are known to have sophisticated mechanisms for protecting themselves against predators, but it is also becoming known that they possess innate defenses against disease.

A newspaper article reported an interesting observation made by an unnamed scientist who was studying the North American tree porcupine when he (or she) was pierced by one of the animal's spines.[1] It was a deep penetration but did not lead to an infection. Could it be that these spines are somehow made sterile? Porcupines are clumsy animals prone to falling and therefore subject to being pierced by their own spines. To explore the possibility that a mechanism existed for protecting them in such a scenario, a study was made of these stiff bristles. It showed that they are imbedded in a thick grease which in fact has bactericidal properties. So, even if a porcupine falls and is pierced with its own spines, there is a reason for it not to become infected.

A similar finding based upon acute observation has been that frogs are blessed with a naturally occurring antibiotic that protects them from infection. Michael Zasloff, at the National Institutes of Health, was using frogs as test animals and noted that they seemed to be immune to infections that might have been anticipated. He was studying the expression of RNA in multicellular organisms by injecting genetic material into frogs' eggs. This required that an incision be made in the frog, and following the closing of the incision, the frogs would be put back in their aquarium. Zasloff noted that "... it struck me that these frogs heal without any inflammation, without pus, or signs of infection."

The aquarium that Zasloff used was often murky and undoubtedly contained high bacterial counts, therefore should have been an abundant source of infection. Zasloff made extracts of frogs' skin and found that they contain at least five major components with antibacterial activity. Upon isolating the two with the highest specific activity, he found them to be closely related and named them "magainins" after the Hebrew word for shield. They were found to be effective against bacteria such as *Escherichia coli, Staphylococcus, Streptococcus, Enterobacter, Pseudomonas,* and also *Candida albicans,* a common yeast that produces infections. Particularly noteworthy was their marked effect on protozoa; within minutes after exposure to a concentration of 10 mcg/ml, paramecia swell and then burst. Incidentally, Zasloff's co-workers had chided him for busying himself with something that was clearly not his primary research goal (par for the course). According to a subsequent interview Zasloff now holds the position of president of the Magainin Research Institute and is enthusiastic about the implications of the current research.[2] Magainins are relatively small molecules, linear peptides roughly twenty amino acids in length; every peptide found in the skin of the *Xenopus* frog has been found in the central or peripheral nervous system of man. What their role can be is speculative at the moment, but surely of interest.

An alert observation by James A. Nathanson and co-workers at the Department of Neurology at the Massachusetts General Hospital (Boston, Massachusetts) led to an interesting finding. It had been found that coca plants tended to be relatively pest free. Cocaine occurs in coca plants, therefore the possible link between cocaine and insects was worth investigating. This led to the finding that insect larvae exposed to cocaine-sprayed leaves display "marked behavioral abnormalities, including rearing, tremors, and walk-off activities." Data developed by Nathanson's group indicate that cocaine's toxicity stems from its blockage of the re-uptake of octopamine, a key insect transmitter and hormone that regulates movement, behavior, and metabolism. The researchers point out that cocaine's effects on humans, caused by blockage of the re-uptake of dopamine (a neurotransmitter), are likely an unintended evolutionary side effect of its ability to block amine re-uptake in insects.

Pigs and Anesthesia

A complication associated with general anesthesia is the fatal syndrome of anesthetic-induced malignant hyperthermia. It is a rare occurrence, but unpredictable, and the mortality rate is seventy to eighty percent. Thanks to a chance occurrence and an astute observation, progress has been made in controlling the problem.

G. G. Harrison tells the story of an interdisciplinary group that included surgeons, anesthetists, physicians, and a laboratory technician, all members of the Liver Research Group of the University of Cape Town, South Africa.[3] They were engaged in a program of experimental liver transplantation using the pig. What they found was that in a particular lot of pigs, six of the first thirty-four animals anesthetized had this reaction. Much is still to be learned of the phenomenon, but knowledge developed from worldwide studies of the malignant hyperthermia swine has led to a pharmacological control of the syndrome and valuable spin-offs in many biomedical areas, particularly muscle and membrane physiology.

Harrison also made the observation in his *Citation Classic* that all the authors of this paper, with the exception of the lab technician, became professors and one became a university vice-chancellor.

An Insight into Endocrine Disruption

Readers of this book will probably know that the low reproductive rate of many birds was the consequence of the reduced thickness of the birds' egg shells. An article by Bette Hileman in *Chemical and Engineering News* indicates that the pernicious effect of DDT (or its degradation products) is still a factor to be reckoned with.[4] In about 1988 the State of Florida asked Louis J. Guillette, a reproductive endocrinologist from the University of Florida (Gainesville), to study the hatchability of alligator eggs to determine how many can be harvested from the wild without threatening native populations. Guillette and his collaborators collected fifty thousand eggs from six lakes and discovered that the hatching rate on most lakes was about seventy percent. On Lake Apopka it was a different story, with only a twenty percent hatching rate. In addition, they found that about half of those that did hatch from this lake died within two weeks. Male alligator hatchlings from Apopka have variously reduced penises, produce elevated levels of estrogen, and fail to produce normal amounts of testosterone. On the other hand, the females produce too much estrogen and their ovaries contain abnormal eggs. Further study showed that DDE, a metabolite of DDT, occurred in these eggs at the high concentration of six parts per million even though the compound was undetectable in the water of Lake Apopka. Probably the whole problem stemmed from a large spill of dicofol in the lake in 1980, and dicofol contained fifteen percent DDT. It appears as if the State of Florida got its money's worth in this study.

Mosquito Sterility

Florida has alligators which it values, and mosquitoes which it hates. One advance toward the possible control of mosquitoes resulted from a sharp observation by a technician at a University of Florida entomology laboratory. When a naturally occurring hormone is given to an adult mosquito by injection or in its food, it cannot make trypsin, which is necessary to digest blood for the maturation of its eggs. Carol Smoyer noticed that treated mosquitoes had an unusually large amount of undigested blood in their gut. Further study showed that a particular fraction of mosquito ovary extract inhibits egg maturation, making the mosquito sterile. It is hoped that this discovery will provide a new type of pesticide.

A Clue to Newcastle Disease

A virus disease causing respiratory and nervous systems in birds, especially domestic fowl, is known as Newcastle disease.

In a report written by S. B. Hitchner describing his discovery of a vaccine for controlling the Newcastle disease virus in chickens, he suggests that researchers who are friends should work in different fields if they want to remain friends. This was the result of a falling-out between him and his former professor, Fred Beaudette (who figured prominently in the narrative on streptomycin in chapter 5). Hitchner had studied under Beaudette at Rutgers University (Brunswick, New Jersey) and had worked in his laboratory as a part-time assistant. It was partly through Beaudette's influence that Hitchner was admitted to the University of Pennsylvania Medical School of Veterinary Medicine. While he was attending veterinary school, he spent his summers at Beaudette's laboratory and even became well acquainted with his family.

After graduation and a stint in the Army, Hitchner obtained a position at the Virginia Agricultural Station of the Virginia Polytechnic Institute (Blacksburg, Virginia). At that time Newcastle disease was prevalent in poultry flocks throughout the United States, and concern by the U.S. Department of Agriculture was sufficient in 1947 to provide research funds for studies of the disease. Hitchner obtained a grant and wrote to Beaudette, requesting samples of appropriate strains of Newcastle disease virus. He also asked for a strain of infectious bronchitis virus (B-1). From what was known, the Newcastle disease strains should give a positive result with a hemagglutination test; the bronchitis virus should not. Hitchner's procedure was to show the contrast. It is sufficient for the purpose here to state that when Hitchner mixed red blood cells from chickens with the eight samples of Newcastle disease virus, there was a hemagglutination response. The same response, however, occurred with the bronchitis virus, and that was not consistent with the premise of the study. At that stage it was thought that all strains of Newcastle disease were highly lethal for young chicks, so if this B-1 strain was a Newcastle virus, it should cause signs to appear in the central nervous system and be fatal to baby chicks. As proof of that premise, a small group of day-old chicks was treated with a drop of the infected embryo fluids of the B-1 strain into each nostril. Upon observation for nineteen days, however, no nervous signs developed nor were there any mortalities. This was indeed strange because it was totally contrary to what had been expected. In an effort to explain what had happened, one of the Newcastle strains was used to inoculate the group that had survived the exposure to the Newcastle virus, and it was also used to inoculate a control group. The group that had previously been inoculated with the B-1 strain showed no effect, but the control group began to die off quickly. Subsequent experiments confirmed the finding that by inoculating the chickens first with the bronchitis (B-1) virus, the chickens were immune to the Newcastle disease.

This most welcome finding was a thrill for young Hitchner, and it was a pleasure for him to inform Beaudette of his results. Unfortunately, Beaudette took the position that Hitchner had made a mistake in his procedure, and he stubbornly refused to admit the possibility that Hitchner's findings were valid. Beaudette was fostering the use of a different vaccine, which would explain part of his antipathy toward the accidental discovery made by Hitchner. What followed was a sordid story of severely strained relations between the two that were resolved somewhat before Beaudette's death in 1957. Hitchner's findings were endorsed by the poultry industry and by many commercial vaccine laboratories.

The Nobel Prize and the Portuguese Man of War

One of the great discoveries early in the twentieth century earned a Nobel Prize for Charles Richet.[5] It concerned the basis of anaphylactic shock and occurred in a roundabout way. The Prince of Monaco commissioned Richet and his collaborators to study the marine siphonophore, the Portuguese man-of-war *Physalia spp.,* and provided them with working space on his yacht during a cruise. They captured a number of specimens and made extracts, but time ran out before any definite conclusions could be made. To continue their work back on land, they switched to sea anemones as their test organism and made various extracts with glycerine. To determine whether there might be any effect of these extracts they inoculated dogs, allowing three to four days after injection before making any judgment about their effects. If a dog survived without any apparent effects, it was used again for another test. During these secondary tests, Richet and his collaborators noticed that there were instances in which dogs died suddenly even though the dosages applied were considered comparatively slight. In studying the records of these test dogs, they concluded that their deaths resulted from a prior sensitization, and thus there came about an understanding of anaphylactic shock. In retrospect, this is another example of a major finding resulting from a totally unexpected routine. Probably no dog ever saw a Portuguese man-of-war, but both played indispensable roles in this Nobel Prize-winning study.

Some Amazing Attributes of Animals

It isn't likely that the reader will need to know how to tell the age of a lizard but, then, you never know. Robert Holmes, of the Department of Surgery at Alfred Hospital in Prahran, Victoria, Australia, serendipitously came upon a way to determine the age of the lizard *Tiligu rugosa.* It is known as a bobtail. Holmes and Alan Light, of the Department of Physiology at the University of Western Australia (Perth), were conducting research on the workings of the ear. They routinely removed parasites from the ear of any lizard they were working on. Commonly a few layers of previously sloughed skin remained, but in the case of one animal, they found both ears completely blocked by compacted skin layers. Further investigation revealed that sloughed skin could not have been removed naturally unless it had disintegrated. When examining one of the plugs from an animal's ear they were able to distinguish 114 distinct layers before subsequent layers began to crumble. By growing the animals in a pen and making periodic studies of the skin structure of the plugs formed in their ears, they were able to tell with certainty the ages of all the test animals. It was found that large animals of this species live to be well in excess of twenty years, something that had not been known.

Stress and Gestation

Information concerning the possible effect of stress on embryos came accidentally from a study of armadillos. Seven female armadillos were captured in Florida in November 1984 and shipped to a laboratory in London. In March 1986 five of the armadillos gave birth. In each case, as is usual for armadillos, each mother gave birth to genetically identical quadruplets. What made the births remarkable was that they came sixteen months after their last possible contact with males, and twenty months after their last mating season in the wild. Armadillos, like a few other mammals, have long been known to delay for three to four months the implantation of early embryos into the uterine wall. This, plus a five-month period of active development, makes their normal gestation period eight to nine months. Curiously, the London armadillos had at least a sixteen-month gestation period. Stress may have caused the delay in gestation because many mammals are known to abort pregnancies in response to stress. The long airplane flight from Florida to London after their capture would surely be considered a stressful situation.

Discovery of a Frog's Capability by Means of a Laser

Laboratory accidents can take many forms and sometimes produce totally unforeseen results. One is related to the discovery of an unusual way in which frogs can hear. At West Germany's Konstanz University, Peter Narins (University of California at Los Angeles) wanted to know why coqui frogs, whose extremely loud calls have sound pressure levels as high as one hundred decibels one meter away, don't deafen themselves when they blast out mating calls. Narins collected the loudest frogs he could find in Puerto Rico and brought them to colleagues Gunther Ehret and Jurgen Tautz, who used laser beams to measure eardrum vibrations. In unfamiliar surroundings, however, the frogs became shy and would not blast out their call, possibly the result of stress. In frustration, and attempting to salvage something from their experiments, the researchers decided to measure the eardrum movements of the frogs as they listened to coqui calls which had been recorded in the wild. Narins recalled, "It was all pretty boring because we were getting just what we expected." But late one night Narins's hand slipped and allowed the laser beam to move to the side of a frog over one lung, which would have been expected to be impervious to sound. Instead, this spot vibrated. In amazement, they mapped the whole frog to find vibrating places; it turned out that only one localized region, about twenty-five millimeters square, vibrates. This was the spot the laser happened to slip to that night.

A Beneficial Finding with a Mouse Tumor

A serendipitous discovery regarding the use of skin grafts for burn victims came about as a result of a study of mouse tumors. Howard Green at Harvard found in these tumors flourishing colonies of cells that resembled those of the upper layer of living skin. These epithelial cells grew because of the presence of fibroblasts, a type of cell common to the connective tissue that comprises the dermis. Green recognized the implication for burn patients because if one could culture skin that was taken from the burn victim, it should not be rejected subsequently by the body's immune system. His procedure was to culture the skin in a flask, and then to remove the skin from the flask surface by means of a bacterial enzyme. Whole sheets of skin could be prepared in this way and transferred to the burn victim. Two young boys who were burned in 1983 with scorched skin over ninety-seven percent of their bodies were saved by this procedure. It is hard to imagine that their survival would begin with the study of a mouse tumor.

The Extraordinary Attraction of a Sex Pheromone

The man responsible for much of the research on the sex attractant of the gypsy moth is Morton Beroza, who retired from the U.S. Department of Agriculture in 1972. He had isolated the components of the sex attractant and had synthesized them, so one would expect a trace residue to be associated with his person or with his clothes. No one could anticipate such a detectable residue many years later, but that is a fact. I had occasion to speak with Beroza in about 1987, which would be fifteen years after his last association with the pheromone study, and he told me that when he perspires (he is a jogger), and the gypsy moths are flying, they still congregate around him! That is an unbelievable story except that I have known Morton for many years and can assure the reader of his veracity. Others who worked at the same Beltsville laboratory in a Maryland suburb near Washington DC, have had similar experiences. Perhaps there will come a time when we can devise a system based on this extreme sensory perception for diverse uses.

REFERENCES

1. *Washington Times.* Oct. 27, 1991.

2. Zasloff, Michael. 1994. An interview. *Current Content* (LS). March 14.

3. Harrison, G. G. 1987. *Current Contents.* No. 16, March 9, referring to Harrison, G. G. et al. 1969. Anesthetic-induced malignant hyperpyrexia and a method for its prediction. *Brit. J. Anaesth.* 41: 844–855.

4. Hileman, Bette. 1994. Environmental estrogens linked to reproductive abnormalities, cancer. *Chemical and Engineering News.* Jan. 31, pg 19–23.

5. Richet, C. R. *Nobel Lectures, Physiology or Medicine, 1901–1921.* pg 473–490. Elsevier, Amsterdam: 1967.

CHAPTER 9

FROM THE EARTH TO THE STARS

"The question of common sense is always 'What is it good for?'—a question which would abolish the rose and be answered triumphantly by the cabbage."

—J. R. Lowell

Studies of the stars were possibly the first venture of man into science, yet there would be many who ask, "What good is it?" Those directing the U.S. National Aeronautics and Space Administration program might wish they had a better answer than the one normally given, that the technology spin-offs have benefited everyone. In terms of serendipity, astronomy is sometimes considered the chief beneficiary.

Testing a hypothesis in science often consists in setting a stage with all the props in place and determining what happens when a change in conditions is introduced. That is not done conveniently with astronomy. It presents a unique challenge to an investigator, because the options available for introducing variables are limited; one can't just move Venus a little to the left, for example. Of course, once the paths of the planets and stars become predictable, it is possible to include them in an experimental scenario. With this as background, it is reasonable to imagine that chance may have played a greater role in the advancement of astronomy than in most other sciences. Some astronomers have certainly been willing to accept that premise.

The Fruits of Questionable Film

The discovery of the Shoemaker-Levy comet in the spring of 1993 and the excitement caused by its predicted collision with Jupiter is somewhat representative of serendipity's role in science. The Shoemakers (husband Gene and wife Carolyn) and David Levy spent seven or eight nights a month with a telescope, each pursuing his/her own specialty. Carolyn specialized in the detection of comets and was credited with detecting more than anyone else. Her technique consisted in taking pictures at specific time intervals and then superimposing them in such a way that an object moving at a faster rate than its neighbors would become apparent.

Unfortunately, someone committed the unpardonable sin of leaving open a box of film and exposing it to the light. The team's first impulse was to throw the box away lest the malfunction of the spoiled film would ruin an experiment. On the other hand, there may have been some sheets that remained unharmed, so they held onto the box. On the following night the team was on location, but the weather was not promising, and the prognosis for an evening of sky-watching was poor. Shooting any pictures on such a night would probably result in a waste of film. But they had in hand the questionable sheets of film which might have been good, so why not try them? There was nothing to lose, and it was with that film that they discovered the comet that was to be so famous. Before its subsequent impact with Jupiter (July 16–22, 1994), the comet had broken into nine pieces and thus became known as the Shoemaker-Levy 9. With the whole world alerted to the coming event, this may have been the most reported phenomenon in the history of astronomy. It might not have happened if someone had not goofed in leaving open a box of film.

The Discovery of Radio Astronomy

In 1930 an electronics engineer, Karl Jansky, was hired by Bell Laboratories to investigate the cause of the hiss type of static that hampered radio reception. Jansky thought of three potential causes: thunderstorms, sun spots, auto ignitions (because by this time there had been a significant increase in the number of automobiles).

His first step was to locate the direction from which the static came. He designed a highly directional array that could be pointed at any section of the sky. To provide mobility he mounted it on a frame set on four wheels salvaged from an old Ford. Jansky learned quickly that the static varied greatly with location. He learned further that for a given location the static changed with the time of day and with the season. Jansky amassed a great body of data, which apparently had no order. He happened to discuss the problem with a coworker, George Southworth, whose primary interest had been earth currents (another example of the value of gabfests). Southworth suggested that Jansky plot all of his data as a function of time, since that had a connection with the earth current hypothesis. When this was done there was no apparent relation to earth currents, but the exercise did reveal that the same signal could be found in a cyclic pattern (i.e., for a given location there was an exact repetition of static a year plus one day later).

Jansky found this to be interesting, but it had no significance until he mentioned it during a bridge game. One of the players, also an employee at Bell Labs, was Melvin Skellett, who was studying astronomy in evening classes at Princeton. When Skellett heard of the year-plus-a-day cycle, he immediately connected with his understanding of the annual cycle of the stars and suggested that Jansky's hiss static was a product of the Milky Way. Further studies confirmed the connection.

Two totally unforeseen circumstances aided in that study. That year, 1932, marked the low point in an eleven-year cycle of sunspots, so interference from the sun was minimal. Also, in August of that year there was an eclipse of the sun and Jansky's measurements made before, during, and after the eclipse showed that the sun had no effect on the phenomenon.

To sum up, a study oriented toward improving radio reception resulted in a whole new field of astronomy. It was chance that the study flowered during a time of low sunspot activity; it was chance that an eclipse occurred at an opportune time; and it was chance that Jansky played bridge with someone who had a knowledge of astronomy. All of these elements contributed to the recognition of the new field of radio astronomy.

An Explanation of Solar Flares

It is interesting that so many researchers found their way into astronomy after being trained in other disciplines. Also, there are instances in which phenomena in astronomy might be explained because of experience gained in other fields. Such was the case with solar flares.

In the 1950s it was understood that solar flares might have an electromagnetic origin, and that the flare energy is probably stored in the solar atmosphere. What was not clear was how the stored magnetic energy could be released within the short time observed from the solar atmosphere that was usually considered to have an almost infinite conductivity. To explain solar flares, Hannes Alfven and

Per Carlqvist developed a model that was based on certain problems related to the electrical power supply system in Sweden. It had been found that if the electric current in a mercury rectifier were increased above a certain critical limit, the current through the rectifier was interrupted within a small fraction of a second. As a result of the interruption, the electrical energy of the circuit was concentrated and then released in the rectifier with disastrous consequences. The current interruption was caused by an electrostatic layer of high impedance. This double layer, through some instability, locally replaced the normally well conducting mercury plasma in the rectifier. Alfven and Carlqvist suggested that a similar kind of double layer might arise in current systems penetrating the solar atmosphere. This would then lead to an explosive release of magnetic energy in the form of flares. The authors, asked to give an account of their research as a *Citation Classic,* suggested that if we search any field of science we will find that progress is often delayed, even for decades, by the lack of contact between groups working in different fields. In this case their knowledge of the electrical power supply system in Sweden led to an important discovery in astronomy.

Alfven was the cowinner of the Nobel Prize for physics in 1970, marking a triumphant victory over his detractors for many years. At the time of his death in 1995, accounts were written of the problems he had with his contemporaries, who often regarded him as a heretic. There were instances in which he was forced to publish in journals of low esteem because his views were so far ahead of his time. He played an important role in the development of plasma physics, charged particle beams, and the branch of plasma physics known as magnetohydrodynamics. His wide-ranging interests inevitably caused friction with the established experts in various fields, and thus it was that his discovery of hydromagnetic waves caused ripples. Winning over the opposition generally requires an admission by one of the experts that the newly espoused views have merit, after which the other experts join the line. Alfven's theory of hydrodynamic waves in 1942 conflicted with Maxwell's theory of electromagnetism. Alfven's work was not acknowledged until six years later, according to University of Arizona professor Alex Dessler, former editor of *Geophysical Research Letters.*

> During Alfven's visit he gave a lecture at the University of Chicago, which was attended by Enrico Fermi. As Alfven described his work, Fermi nodded his head and said "Of course." The next day the entire world of physics said, "Oh, of course."

An Insight into Moon Rocks

Circumstances in one's early life can become important with the passage of time. That is surely the case with Edwin Roedder who became prominent in the study of moon rocks, something which would have defied prediction when he began graduate work at Columbia University after World War II. Actually, the chain of events began much earlier when Roedder's mother fostered his interest in mineral collecting by taking him on trips extending from Maryland to New England.

While an undergraduate at Lehigh University (Bethlehem, Pennsylvania), he was impressed by a slide of quartz containing tiny bubbles in fluid inclusions. Roedder was fascinated by the thought that these bubbles might have been bouncing around in their prison cells for hundreds of millions of years. While at Columbia his mentor there, C. H. Behre Jr. made an unusual arrangement whereby Roedder could do his doctoral research under J. Frank Schairer at the Geophysical Laboratory of the Carnegie Institution of Washington. Roedder credits Schairer with making an excellent suggestion that he specialize in a certain system of K_2-MgO-SiO_2. This formed beautiful highly modified hexagonal crystals that melted incongruently upon heating, and which formed a phase called "forsterite plus liquid." Roedder predicted that it should occur in nature, but the fact that its optical properties matched quartz exactly was not recognized. Later it was of great satisfaction to him when the mineral was actually found in meteorites and later in volcanoes, and was given the name "roedderite." His familiarity with liquid inclusions in rocks became most useful at the time our astronauts brought back to Earth a collection of moon rocks. Roedder was uniquely qualified among geologists to identify them because of their similarity to the synthetic compositions he had studied.

It was my great pleasure to know Dr. Schairer when I worked at the Geophysical Laboratory during World War II. He was a prodigious researcher and the story is told that following World War II he had several visitors from the Soviet Union who were familiar with the magnitude of Schairer's publications and had expected to find someone with a large staff actually performing the work. They were amazed that he would repeatedly excuse himself from their conversation to take a reading on a thermocouple. He was a one-man show in the best sense of the word.

My most treasured remembrance of Schairer was that of an unparalleled storyteller. Those who had the privilege of hearing him describe the unscheduled stops of a train heading for his favorite fishing spots in Canada during the days of Prohibition would not forget him. The cognoscenti aboard the train during these unscheduled stops would head for the moonshine stills known to exist in the forests. At the end of this chapter is Schairer's description of a chance meeting with several fellow hikers on the Appalachian trail.[1] The scientific world lost one of its most beloved when Dr. Schairer, renowned scientist, orchid fancier, and raconteur drowned accidentally.

Hanbury Brown, Radio Engineer/Astronomer

An example of someone switching to astronomy from another field is Hanbury Brown, who started his working life as a radio engineer in 1936. He worked on the early development of radar for the British Air Ministry, and after the war joined Sir Robert Watson-Watt as a consulting engineer in London. In 1949 Sir Robert moved the firm to Canada, and Brown decided to stay and look for another job. A professor in the Electrical Engineering Department at the University of Manchester mentioned to him that Bernard Lovell was doing some work at Jodrell Bank. When Brown visited Lovell he found that the group there had built a large parabaloid antenna and with it one of the students was trying to detect cosmic rays. Lovell felt that Brown's experience as a radio engineer would be useful, so he invited him to join the team. The result was that Brown worked on astronomy from then on.

Serendipity played a role in his accomplishments when he was working on a time correlation between photons and the subsequent development of the intensity interferometer:

> It was while Richard Twiss and I were watching our radio intensity interferometer working in 1952 that we just happened to notice that the fluctuations due to the ionosphere did not appear to affect the results. We began to wonder why this should be so, and the answer led us to the idea of applying the same principle to optics, and that in turn led us to the idea that photons in mutually coherent beams of light must be correlated in time—an idea which aroused a great deal of quite passionate opposition but which turned out to be right.

A Non-Stellar Companion to a Sun-Like Star

According to an article by M. Mitchell Waldrop, serendipity played a role in the discovery of an apparently unremarkable star that turned out to be most interesting.[2] This star, HD 114762 is about ninety light years from earth and too dim to be seen by the naked eye. Its spectrum indicates that it has only one-tenth the abundance of heavy metals that the sun has, from which it is inferred that it is five to ten billion years older than the sun. Other than that, it is almost identical to the sun.

Team leader David W. Latham of the Smithsonian Astrophysical Observatory reports that they stumbled upon this star, which appeared to have a non-stellar companion—a "planet" that paradoxically combines a mass at least ten times that of Jupiter with an orbital radius as small as that of Mercury. Latham's group looked for sub-stellar companions by checking hundreds of stars and looking for subtle but periodic shifts in their spectra. The basis for this is that an unseen companion can tug a star back and forth and create a Doppler shift in the star's spectral lines. One step in this process is to settle on some standard constant-velocity stars that can be used to calibrate Doppler shifts on any given observation run. After many measurements, they settled on nine stars as their standards, one of them being the unremarkable HD 114762. However, as time went on they became aware that this star was not a model of stability, but rather that its spectra showed periodic variations in Doppler shift of something like five hundred meters per second. Mathematical techniques developed by Tsevi Mazeh (Tel Aviv University) deduced from the data a periodicity of eighty-four days. Assuming that the signal was real, this would give the companion star an orbital radius comparable to that of Mercury, which orbits our own sun every eighty-eight days. An independent confirmation of the findings came from Geneva Observatory's Michael Mayor, who used a different instrument and found the same orbit.

New Technique Finds Dim Galaxies

An article by Ivars Peterson describes a serendipitous finding of hard-to-detect dim galaxies.[3] To identify a dim object against a background of natural faint light from the night sky would obviously be a difficult task, but it was necessary in order to find elliptical "dwarf" galaxies. These are common, but as members of galaxy clusters they are particularly hard to find. An ingenious method for this purpose was devised by David F. Malin of the Anglo-Australian Observatory in Epping, Australia. Based on the fact that very faint images are recorded only by the uppermost layers of a photographic emulsion where the grains of silver forming these images are concentrated, Malin can amplify the images by passing diffuse light through the emulsion onto a piece of high-contrast film. This greatly increases the apparent size of the original grains. Malin and his associates Gregory D. Bothun (University of Michigan at Ann Arbor) and Christopher D. Impey (University of Arizona at Tucson) used this method to study a galaxy cluster in the direction of the Constellation Virgo. This cluster is the nearest large collection of galaxies to the Milky Way. They identified twenty-seven new dwarf galaxies, including one that had two distinctive features—a slightly fuzzy nucleus, and a faint outer envelope showing traces of a pattern. Subsequently it was found that this unusual object is the largest galaxy yet discovered. It lies about 715 million light years from Earth and has a diameter of at least 770,000 light years (the diameter of the Milky Way is only one hundred thousand light years). Since then, this team has discovered another such faint galaxy but it is only half as large.

Supernova 1987A, World Class Event

No supernova has been clearly visible to the naked eye since 1604. That one was detected four years before the invention of the telescope and was observed extensively by Johannes Kepler. So the event itself (the death of a massive star) is unusual, but circumstances set this 1987 event apart from all others.

First, it occurred in a satellite galaxy of our own, in the Large Magellanic Cloud, and while that is 160,000 light years away, it is relatively close compared to others that have been known to occur. Second, it was possible to observe its effects by high technology techniques that had not been available until recently. Third, the role played by amateur astronomers was critical to an interpretation of the phenomena that resulted.

Ian Shelton, a Canadian astronomer, announced his discovery of Supernova 1987A on the night of February 24, 1987. Fortunately, Shelton had taken a picture of the same field the night before his discovery when there was no evidence of the supernova, so light from it had appeared only within the past twenty-four hours.

Serendipitously, R. H. McNaught had recorded the supernova only eight hours after Shelton's plate of February 23. McNaught's photograph was taken only three hours after the detection of a neutrino burst had been made by Kamiokande with a deep underground detector. McNaught was disappointed that he had not examined his photograph immediately and therefore been credited with the discovery. A New Zealand amateur astronomer, A. Jones, was scanning the appropriate region of the sky with a small telescope on the night of February 23, but did not notice the supernova. This suggested that it was only one-third as bright as when McNaught photographed it seventy-eight minutes later. Although the discovery was credited to Shelton, the data points in the light curve supplied by McNaught and Jones were valuable in the construction of hydrodynamic models.

An easily readable account of the findings by S. E. Woosley and M. M. Phillips was in *Science.*[4] Woosley is professor of astronomy and physics at the University of California, Santa Cruz, and Phillips is associate astronomer at the Cerro

Toloko Inter-American Observatory in Chile. I wrote to Woosley and told him of the pertinence of Supernova 1987A to serendipity, which I was to incorporate into this book. In his reply, he pointed out that serendipity played a role in three out of the four Nobel Prizes that have been awarded in astronomy and astrophysics:

> Hans Bethe won the first in 1967 for work he did in the 1930s explaining how energy is produced in the sun. This was not accidental, but the result of considered thought and reason. However, the next prize was for the discovery of the three degree (Kelvin) microwave radiation background by Penzias and Wilson in 1965. This is the piece of evidence that convinced us all of the Big Bang. They weren't looking for it, though unbeknownst to them it had been predicted. They just had a source of background in their detector that wouldn't go away. Next came a prize to Hewish for discovering pulsars. Actually his graduate student, Jocelyn Bell, made the discovery, a 'bit of scruff' in the radio signal from part of the sky that repeated from time to time. It was noise, she looked for ways to get rid of it, but it stayed. Last, in 1984, came the prize jointly to Fowler and Chandrasekhar, the former for working out the production of heavy elements in the stars, the latter for the *Chandrasekhar mass* (and other things), the upper limit to the mass of a white dwarf star beyond which it must collapse to a neutron star or black hole. The work of Fowler was in the nuclear laboratory, not serendipity, but Chandra wanted to understand better the physical structure of white dwarf stars. When he put corrections for special relativity into his equations, he got a result that surprised him greatly. Above 1.44 times the mass of the sun, no stable white dwarf could exist. All the exciting physics of gravitationally collapsed objects—black holes and neutron stars—followed.

Woosley mentioned several other astronomical findings that were the result of serendipity. In one instance astronomers noted a few unusual faint blue stars that had spectra that no one could identify. Maarten Schmidt realized that the spectra represented ordinary hydrogen, which had an extremely large shift toward the red end, and they were at a great distance. So far away were the objects that their luminosity, to be detected, must have been tremendous. The explanation was that these were quasars.

A second instance cited by Woosley is the discovery of gamma ray bursts. Although they are still not understood, they were discovered by the VELA satellite network which was set up by the Department of Defense to monitor the Earth for clandestine nuclear tests.

Woosley ended his most welcome letter with an interesting statement about serendipity:

So time-honored has serendipity become in astronomy that it is a major consideration at NASA when discussing prospective new space instruments. One can be pretty sure that whenever a new band of wavelengths or a new lower level of sensitivity or a new range of temporal variability are explored, that new and exciting phenomena will be discovered. They are right. It may be serendipity, but if you don't look you will never see.

Luck, Amateur Style

As we have seen, amateurs have helped establish the sequence of events surrounding Supernova 1987A, and there was the factor of luck concerning the man being given credit for the discovery. But one of the most improbable astronomical discoveries had to be that of Ben Mayer, an amateur astronomer in Los Angeles who died in 1999. A retired businessman, Mayer's long interest in astronomy has resulted in several books on the subject. His base of operations was not ideal for studies of the stars—the roof of his residence in Los Angeles, that smog-laden city with many bright lights. Mayer had a camera attached to his telescope and was able to take sequential pictures with the aid of a timer he had rescued from a lawn sprinkler. He happened to have the telescope aimed in a favorable direction one night, and the resulting sequence of pictures he obtained showed the birth of a nova that became the subject of many studies throughout the world.

Saturn's Outer Ring

Daring to do something differently can produce interesting results. A case in point is that of Walter Feibelman, a research engineer in electronic imaging at the Westinghouse Corporation (Pittsburgh, Pennsylvania), who served as a part-time observer at the University of Pittsburgh's Allegheny Laboratory. For the type of work he was doing, the practice had been to expose photographic plates for periods ranging from seconds to a maximum of a minute. Feibelman went far beyond that limit and used exposures of five to thirty minutes, the result being the discovery that Saturn had an additional ring that had been unknown. Naturally, this caused a stir among astronomers whose devotion to the conventional wisdom had been challenged, but the proof came with NASA's *Voyager II* which came close enough to the planet to certify the existence of the extra ring.

Separating Two Stars by Chance

About thirty years ago George Wallerstein, professor of astronomy at the University of Washington (Seattle), was using the Lick Observatory's 120-inch telescope one night, but daylight was approaching and he had only ten minutes of darkness remaining. He was working at a high spectral dispersion and was limited to viewing only very bright stars in the time he had left. One of the elements of interest at that time was the depth to which the outer layers of stars were mixed by convection. A tool for studying this was to look for lithium in the stellar atmosphere. Nuclear reactions between lithium and hydrogen begin at about 2×10^6 degrees Kelvin for ^{6}Li and, 3×10^6 degrees for ^{7}Li. The lithium is rapidly consumed. Wallerstein spotted Capella, a first-magnitude binary star that had never been resolved visually, but he obtained a high-resolution spectrum of Capella and found that the hotter of the two stars showed a strong lithium line whereas the cooler component did not.

Discovering a Satellite of Pluto

It was hard enough to discover the planet Pluto in the first place, so the prospect of detecting a satellite around it had to be slim. The search for the planet began in 1906 by Percival Lowell, who became interested in the possibility of an unseen planet beyond Neptune. From 1906 to 1916 Pluto was moving on the more remote part of its orbit and its brightness was about half of what it was when it was officially discovered. Perturbations in the motion of Neptune could be explained on the basis of a gravitational pull of another planet. It was not until 1929 that a thirteen-inch photographic telescope with a wide field became available, making it possible to search a wide area of the sky at one time. With skillful modifications of special holders for the photographic plates, which bent them slightly into a concave shape, it was possible to produce images less than 1/30th of a millimeter in diameter of the most faint stars. From then on there was the tedium of making multiple photographs of the sky and studying them; the plates taken of the western part of Gemini contained over three hundred thousand star images each. By the latter part of February 1930 there was sufficient evidence for Clyde Tombaugh to show that there was a planet, which became known as Pluto.

Forty-eight years after that in 1978 a regular series of observations made by the U.S. Naval Observatory telescope at Flagstaff, Arizona, was compared with photographs taken at its facility in Washington, DC. While measuring plates taken of Pluto on three nights in April and May on the STARSCAN automatic measuring machine, James Christy noticed that the images appeared to be consistently elongated and the instrument seemed to be malfunctioning. While waiting for the repairman to arrive for a look at the instrument, Christy took a closer look at the plates and became aware of the possibility that the elongation was real and not an artifact. When he looked through the archives for other pictures of Pluto he saw the same slight discrepancy, ultimately attributed to the presence of a moon. It was the need to wait for the repairman that prompted Christy's second look at the photographic plate and that, in turn, led to the discovery of the moon of Pluto.

Archival photographs were responsible for another discovery, by B. Schaefer of MIT, also in 1978. The location of a burst of gamma rays coincided exactly with a starlike image that he found in an archived photograph made fifty years earlier in 1928. Schaefer has since made several similar findings.

Accidents Aren't Always Bad

Geologic faults have been reported to release electromagnetic signals prior to generating large earthquakes. Antony Fraser-Smith of Stanford University (Palo Alto, California) contributed to this body of thought in an unusual way. He had been working with instruments that detect very-low-frequency radio waves, and one of his machines detected the unusual magnetic disturbance before the Loma Prieta earthquake in October 1989. With that unplanned discovery as a potentially fruitful lead, he set up two receivers over the San Andreas fault near the town of Parkfield, which has been considered by seismologists as the site of a future large quake. Also, he placed instruments near the San Andreas fault in southern California and in the Bay area. According to an account in *Science News* one of the Parkfield instruments started showing electromagnetic pulses ten times the size of typical atmospheric signals in November 1994.[5] About the same time, seismometers and other types of monitoring devices at Parkfield began picking up unusual signals. A week later, a magnitude 5.0 earthquake struck. To counter this fortuitous discovery, his instruments in southern California had not picked up changes before the Northridge earthquake in January 1994. These results show that more must be learned about the utility of ultra-low-frequency radio receivers in detecting incipient earthquakes.

High Frequency Radio Emissions as Monitors for Earthquakes

According to the same *Science News* issues, astronomer K. Maeda of the Hyogo College of Medicine in Nishinomiya, Japan, had been studying high-frequency emissions from Jupiter, but in the predawn hours of January 1995, his antennas picked up pulses that could not have come from Jupiter. Just forty minutes later, a 6.9 magnitude earthquake devastated the town of Kobe, Japan, just seventy-seven kilometers from his observatory.

Maeda combined two pieces of evidence to link the radio emissions with the earthquake: signals from Jupiter vary continually because of the earth's rotation, but the signals on January 17 were constant, and the emissions came from the general direction of the earth's epicenter on the Nojima fault.

So, two leads have been developed linking radio emissions with earthquakes, one based on low frequencies and one based on high frequencies. And neither one had been planned with the study of earthquakes in mind.

Continents and Magnetism

Edward (Ted) Irving of the Pacific Geoscience Center, Geological Survey of Canada (Sidney, British Columbia) was one of the members of the Royal Society responding to the inquiry I had sent. He supplied a reprint of his paper "The Paleomagnetic Confirmation of Continental Drift," which touched on several recurring themes in this book. He cites examples where prevailing wisdom held sway for many years despite evidence to the contrary. One of the circumstances that helped forge a new concept was the availability of a special magnetometer that had been designed for a different purpose, but was admirably suited for studies of paleomagnetism (the intensity and direction of residual magnetization in ancient rocks).

The magnetometer had been designed years earlier by P. M. S. Blackett, professor of physics at the University of Manchester (UK), who had become interested in the possibility that magnetism might be a fundamental property of rotating bodies—the "distributed" theory of the earth's magnetism. It would suggest that there was a relationship between the angular momentum of matter and magnetism, which was a consequence of the knowledge that stars had magnetic fields. Blackett's elegant magnetometer, however, did not develop any detectable magnetism as a consequence of rotating with the earth each day, as it should have done if the "distributed" theory were correct. But the fact that it did serve an indispensable purpose in paleomagnetic studies is one more example of the unpredictability of applications of science; an instrument spawned by astronomical observations serving the needs of someone working with rocks. Research performed by Ted Irving and others mentioned in the narrative played an important role in the ultimate recognition that the continents do drift.

* * * *

The following is a copy of part of *Biographical Memoirs* Volume 66, a *Biographical Memoir of J. Frank Schairer (1904–1970)* by H. B. Yoder Jr. The kind permission of the National Academy of Sciences was given for publication in this book:

> We started up the hollow and passed the time with everybody. Everybody was friendly but distant. After all, we were foreigners—anybody who lives more than two miles away is a foreigner. Then we started out on an old road. It got narrower and more and more gullied. Finally, it was just a mountain trail.
>
> All of a sudden around a sharp bend in the trail came two mountaineers. One was an older man with a white beard, and the other was a younger man carrying a gunnysack in which it was obvious there were four two-quart jars of corn liquor.
>
> So we just sat there. And there was an awkward pause. And then the conversation got going as they do in the great circle of the mountains—it was a hell of a fine day, or damned if it wasn't. You start with the weather and you end with the weather.
>
> And the next thing you talk about is the crops, which are important to the mountain people, for if the crops are bad they might starve.
>
> And then the talk was about illness, the miseries, as they called it. And about that time, everybody was sick, with inadequate food and inadequate housing, and so forth.
>
> And then another adequate topic of conversation was this proposed Shenandoah National Park, was that all nonsense or was it going through. And we said, "Yes, they are going through, and they might make their plans accordingly."
>
> And we got back to the weather, if it was a good day it was a good day for a drink. Or if it was a bad day, we need a drink.
>
> And the fellow says, "Do you fellows ever drink?"
>
> And I said "I don't mind if I do."
>
> And he brought out a two-quart fruit jar.
>
> Charlie is a nice guy, but he doesn't drink. It was the most embarrassing thing in the world. I rushed up to Charlie and grabbed the two-quart fruit jar. I nearly knocked him down. I swung the fruit jar up, took a good swig, and swung it down again, and I said , "Charlie doesn't drink, but I drink for him." And I took another good swig.

So they thought it was so cute I got Charlie's drink.

And there was an awkward pause. And it suddenly dawned on me that I had a drink in my pack. We never drank on the trail, because climbing mountains and drinking liquor didn't go too well together, but Sunday night when we got to Mrs. Meig's, before dinner a right good snort was always appreciated.

And so we weren't carrying a two-quart fruit jar, but I had a pint thermos bottle filled with liquor in my pack. And I said, "Won't you have a drink of my liquor?" And I went over to get it out of my pack.

The mountaineers can't figure what anybody would put in those packs.

So I pulled out my raincoat and my flashlight and what was left of a sandwich, and the liquor wasn't there; it was in the side pocket. So I pulled out everything trying to find it. And finally I pulled out this pint thermos bottle and handed it to the fellow.

And he took a drink and he looked very startled. He took another little drink, and he handed it back. I put it back in the pack and tied up the pack and we sat down and there was an awkward pause.

Then the old fellow says, "I can tell where you all got that liquor." And I said, "You can?" And he said, "Yes, that is Hazel Hollow liquor."

And Hazel Hollow was about thirty or forty miles to the north.

And I said, "Yes?"

And he said, "I can tell you who made that liquor."

And I said, "Can you?"

And he said, "That is Jack Dodson's liquor. And I can tell you when you got that liquor."

"How can you tell me that? When I got it, it was in a two-quart jar."

But he was right all the time. And he looks at me and said, "You must be Mr. Frank."

Schairer was too much for them; they all called me Mr. Frank.

Here I give a guy a drink of liquor, and he tells me my name. And I said, "Would you mind telling me how you do it?"

He said, "Each hollow has its own formula. There is only one make of liquor in Hazel Hollow, and this is Jack's. And Jack has only made three batches this year. The first batch, he was terribly thirsty, so he let the batch burn. It couldn't have been that, because it was burned. And having burned the first batch he was terribly cautious, and the second batch was perfect."

That was the batch I had. In fact, it was so good that word got around and it only lasted three days. So he knew within three days of when I bought it.

It couldn't have been that third batch, because he had stored the third batch.

The first batch was burned, the second was that good batch, and it only lasted three days and I had it.

"And Jack never sells any of his liquor to anybody outside the mountains but this fellow Frank, and so you must be Mr. Frank."

REFERENCES

1. Yoder, H.S. Jr. Biographical Memoirs. National Academy of Sciences. Vol. 66.

2. Waldrop, Mitchell. 1988. Secrets of an unremarkable star. *Science.* 241: 790.

3. Peterson, Ivars. 1987. The dregs of the universe. *Science News.* 136: 60.

4. Woosley, S. R. and Phillips, M. M. 1988. Supernova 1987A. *Science.* 240: 750–759.

5. Monastersky, Richard. 1995. Radio hints precede a small U.S. quake. *Science News.* 148: 431.

Chapter 10

Winning the Big One

"My colleagues Melvin Schwartz and Jack Steinberger join me to express our feelings of pleasure and gratitude for the decision to award us the Nobel Prize in physics, thus making us experts on the Brazilian debt, women's fashions, and social security."

—Leon Lederman

Prestigious awards are abundant in research but none provide the recognition that is afforded by the Nobel. Leon Lederman's jesting remark illustrates the esteem reserved for those who have won this cherished award. At a time when athletes, politicians, and entertainers are often derided for their lack of intelligence, Nobel Prize winners are left unscathed.

It would seem heretical to suggest that chance or luck could play a role in an accomplishment that culminated in a Nobel Prize, but that is not rare. Unusual circumstances have played important roles in at least nineteen of them, but that does not diminish the merit of the award; being alert to a chance event, and capitalizing on it, sets the exceptional person apart from the merely competent. Albert Szent-Georgi's comment was accurate: "Discovery consists of seeing what everybody has seen, and thinking what nobody has thought."

Accounts of the serendipitous aspects of the discoveries of penicillin and streptomycin are dealt with in separate chapters of this book. Also, the studies of Charles Richet, Albert Szent-Georgi, Alexander Fleming, and A. J. P. Martin are in other areas of this book where they are more pertinent.

Ionizing Radiation

C. T. R. WILSON DONALD GLASER
PATRICK BLACKETT CARL ANDERSON

A chance observation made on a Scottish hillside led to a cascade of discoveries, including several which merited the Nobel Prize in physics. When C. T. R. Wilson won the Nobel Prize for physics in 1927, his address was entitled "On the Cloud Method of Making Visible Ions and the Track of Ionizing Particles." A quote from that address provides the origin of the discovery:

> In September 1894, I spent a few weeks in the observatory which then existed on the summit of Ben Nevis, the highest of the Scottish hills. The wonderful optical phenomena shown when the sun shone on the clouds surrounding the hill-top, and especially the coloured rings surrounding the sun, greatly excited my interest and made me wish to imitate them in the laboratory.

His successful attempt to duplicate at least one facet of the natural scene consisted of designing a closed chamber, with a little water inside, in which he could form droplets by rapidly reducing the pressure. With the addition of a radioactive source to the chamber, the track of a charged particle cutting through the airborne droplets could be recorded by a photographic film, thereby providing a measure of the particle's energy. This apparatus became known as the Wilson Cloud Chamber and provided a means for determining the energy of radioactive decay products. Occasionally a cosmic ray would enter the chamber and create a track of particles having high energies, but the chance of recording such an event was slim.

An improvement on Wilson's Cloud Chamber was devised by Patrick M. S. Blackett in the early 1920s. In 1925 he obtained the first photograph of the disruption of a nitrogen nucleus by an alpha particle. By 1932 Blackett had become interested in cosmic rays but was aware of a significant deficiency inherent in the Wilson Chamber for studying these rays: there was no way to predict when one would appear. Blackett made a major advance by automating the Wilson Chamber such that the cosmic rays would essentially photograph themselves as they passed through the droplets. By mounting Geiger counters above and below the expansion chamber and developing suitable circuitry, the camera recording the event operated only when both Geiger counters were activated. This did not involve luck; it was good science. But there was an unexpected bonus to this approach. It became apparent that some of the electrons being detected had the same mass but opposite charges, occurring in pairs in cosmic rays. Blackett received the Nobel Prize, with Occhialini, for this discovery in 1948.

Wilson and Blackett used photographic tracks associated with the formation of water droplets. With a different approach, Donald Glaser determined the number of bubbles released by high energy particles from a supersaturated liquid. A need to determine the extremely high energies of particles produced by accelerators (several billion electron volts) prompted the discovery for which Glaser was awarded the Nobel Prize in 1960. Legend has it that Glaser was seated in a bar one afternoon when his beer glass happened to be in the path of a sunbeam. Bubbles originated at the point where the sunbeam hit the liquid and he interpreted correctly that the beer, a supersaturated liquid, was releasing the bubbles as the result of the energy in the sunbeam. (I wrote a letter to the address that I had been given for Glaser, to ask about the accuracy of this report, but received no answer). His initial attempts to exploit this phenomenon, using beer and various soft drinks in small chambers, were unsuccessful. The work was expanded to include liquids such as ether, but heated to a high temperature, and liquid hydrogen in chambers almost two meters long, 0.5 m wide, 0.5 meter deep. To establish conditions under which the incipient boiling could be controlled was a major accomplishment which provided a way to measure the extremely high energies of particles produced by accelerators. These are on the order of twenty-five billion volts, more than one thousand times larger than those which were obtainable earlier.

Carl Anderson won the Nobel Prize for physics in 1936, with Viktor Hess, for his studies of the tracks of cosmic rays under the influence of magnetic fields. His work was based on the Wilson Cloud Chamber and led to the unexpected discovery of the positron (positive electron). Therefore, at least four Nobel Prizes originated largely from C. T. R. Wilson's fascination with the clouds seen from a Scottish hillside.

A Payoff of Humor in Cosmic Research

VIKTOR HESS

The humor associated with this account did not concern Viktor Hess, but his Nobel-winning research was a necessary prelude. What prompted his research was that a comparison had been made of the background radiation at the base of the Eiffel Tower with that at the top, and the radiation was found to be greater at the base. Hess explored this phenomenon in detail with a manned balloon flight on April 12, 1912, in which he measured the radiation as a function of altitude. He used a hermetically sealed vessel to measure ionization, and the date was chosen to coincide with an almost total eclipse. His findings were that the eclipse had no effect, and that the radiation decreased until the balloon rose above one thousand meters, but at five thousand meters the radiation was several times greater than it was at ground level. This was the beginning of a series of studies which demonstrated the existence of cosmic rays, for which Hess was awarded the Nobel Prize in 1936.

There was an unusual amount of newspaper coverage of the cosmic ray story in 1936. It was less than that associated with the early NASA flights into space, but it was much greater than what is normally associated with Nobel Prizes. At that time, several scientists announced plans for an extended balloon flight to study the diurnal variation of cosmic rays. An innovative newspaper reporter suggested that this was a dangerous mission on several counts. Primary concerns were the low temperatures to be encountered and the possible fatal consequence of a mechanical failure of the balloon. Also, the press had introduced the conjecture that this extensive exposure to cosmic rays might result in a sex change of the balloonists! With this unusual interest in the mission as background, a practical joker delivered to the launch site a box of sanitary napkins for use by the balloonists in case they underwent a sex change. The sanitary napkins were taken aboard and, believe it or not, fortuitously saved the mission. More atmospheric condensation took place than had been anticipated, but the sanitary napkins absorbed it and thus all of the electronic equipment was saved.

MEDICINE

Viruses

BARUCH BLUMBERG
D. CARLETON GAJDUSEK

In 1976 the Nobel Prize for Physiology or Medicine was awarded jointly to Baruch Blumberg and D. Carleton Gajdusek, though their research was not the result of personal collaboration. Certain similarities marked their research: studies of populations widely scattered over the earth, studies relating to viruses, and results far removed from what had been their initial targets. Some of their research was not planned, could not be foreseen, but was completely necessary.

Baruch Blumberg's award was given for providing a method to detect the hepatitis virus in blood.[1] He had studied at Oxford under A. G. Ogston, and his prize-winning work depended in part upon contributions made by some of Ogston's other pupils. From studies of Basques, Europeans, Nigerians, Alaskans, Eskimos, Indians, Micronesians, and Greeks, he learned of the diversity of biochemical and genetic types to be found in blood serum.

By 1960 Blumberg was at the National Institutes of Health, where he was joined by Anthony Allison, who had been in the Department of Biochemistry at Oxford. They developed the hypothesis that patients might develop antibodies against serum proteins as the result of the blood transfusions they received. For their purposes, someone was "transfused" if that person had at least twenty-five transfusions. Blumberg and Allison found a serum that contained a precipitating antibody in the blood of a patient who had received many transfusions for the treatment of an obscure anemia. Upon repeated testing with this person, they found that the antibody in his blood reacted with inherited antigens specific to proteins with certain fatty characteristics. This they termed the Ag system, and finding it encouraged them to continue their search for other precipitating systems in the serums of transfused patients.

The director of the blood bank at Mt. Sinai Hospital in New York City cooperated by providing the sera of hemophilia patients for study. Antibodies against the Ag proteins were not found often in these sera, but one day they saw a precipitin band unlike any of the Ag precipitins. By the use of staining techniques, they could tell that it was a protein. Most of the antiserums to Ag reacted with twelve to twenty-two of the twenty-four sera in the panel, but the serum from the hemophilia patient reacted with only one of the twenty-four, and that specimen was from an Australian Aborigine. They used the term Au to designate the Australian antigen reactant.

- How could it be that the serum from a hemophilia patient in New York reacted with that of an Aborigine from Australia? The next step was to collect information on the distribution of Au and antibodies to Au in different human populations and disease groups. A collection of serum and plasma samples was established, and this eventually developed into the blood collection of the Division of Clinical Research of the Institute for Cancer Research (Philadelphia, Pennsylvania) numbering more than two hundred thousand specimens. The antigen was very stable and there were instances in which blood that had been frozen for ten years or more still gave strong reactions for Au. Presence or absence of Au appeared to be an inherent characteristic of an individual. From their surveys, the Au was rare in the United States, with only one out of one thousand sera being positive. In Filipinos from Cebu the incidence was sixty out of one thousand, and in certain Pacific Ocean populations it was as high as one hundred fifty out of one thousand. One of Blumberg's associates, Sam Visnich, searched through the sera of patients who had received transfusions to determine whether any reacted with the Au. Some of the transfused sera were from patients who had leukemia, and in this group was a high frequency of Au, rather than antisera to Au. Subsequent testing of patients with other diseases showed that Au was only to be found in those who had been transfused.

Amid this enormous amount of testing there seemed to be several probable relationships:

- Individuals who have Au are more likely to develop leukemia than those who do not have the antigen.
- Individuals who have a high likelihood of developing leukemia would be more likely to have Au.
- Children having Down's syndrome are more likely to develop leukemia than are other children.

All the individuals identified as having Au lived in Australia or some distant place, or they had leukemia. But when Blumberg moved to the Institute for Cancer Research, it was possible to study more people with Au. Having a relative abundance of Down's syndrome patients so close facilitated Blumberg's studies and an important fact emerged; if a patient tested positively for Au initially, it was present later. The converse was also true; if a patient tested negative at first, that patient would remain negative. But in 1966 a Down's syndrome patient who had originally tested negative was found to have Au. This was so unusual that he was admitted to the Clinical Research Unit where a series of tests showed that between the first negative testing for Au and the subsequent positive test, he had developed a chronic form of hepatitis. The diagnosis of hepatitis was confirmed by a biopsy of the liver, and the search was begun for blood samples from other hepatitis patients. It was soon found that many hepatitis patients had Au in their blood early in the disease stage but the antigen usually disappeared from their blood after a few days or weeks.

Then occurred one of those critical events that cleared up the mysteries. A technician who was working on the isolation of Au began to feel ill and tested her own serum for the presence of Au. The following morning the results of the test were evident: She became the first person to be diagnosed for hepatitis by the Au test. By the end of 1966 it had been shown definitely that Au was associated with acute viral hepatitis.

One great consequence of Blumberg's research was that it provided a way to detect the hepatitis virus in blood and is now used routinely by blood banks for this purpose. Estimates are that the annual savings in the United States resulting from this procedure are on the order of a half billion dollars. Even more important is that this procedure has saved countless numbers of people from the terribly debilitating effects of hepatitis.

Carleton Gajdusek, an American pediatrician and virologist, shared the Nobel Prize with Blumberg for his contributions to an understanding of infectious diseases, especially of a chronic or degenerative type. He began his studies in Nigeria in 1957, but when he learned of a disease of the central nervous system in New Guinea, moved there to extend his work.[2] The disease was called kuru, and several of its characteristics were unique. There was no fever or inflammation associated with it, but its victims suffered from increasing weakness and a lack of balance. The disease was ultimately fatal. By living with the natives of the New Guinea highlands and learning of their customs, he was able to associate kuru with a cannibalistic ritual in which the natives would honor their dead by eating their brains. Furthermore, he saw a parallel with kuru and a neurological disorder of sheep, called "scrapie." When the carcass of a sheep was fed to mink, the mink would develop the disease and it would be passed from one generation to another. Because of the lag time involved in the development of kuru, its occurrence had been difficult to track.

The virus responsible for it is included among what are termed "unconventional" viruses which share the following characteristics:

- Resistance to chemicals, such as formaldehyde
- Resistance to enzymes that degrade proteins (e.g. DNA)
- Resistance to temperatures as high as 100° C
- Resistance to ultraviolet light and ionizing radiation

Because of Gajdusek's studies, the ritual cannibalism was discontinued in 1959, and the only cases occurring now are of people who may have participated in that ritual before then. Gajdusek believes there may be a relationship between the etiology of Alzheimer's disease and other neurological disorders such as Creutzfeldt-Jakob Disease, or kuru.

The Middle Ear

ROBERT BARANYI

The human body is enormously complex, and simply learning what organs are present in it can be a large task. An awareness of one organ, the middle ear, can be attributed to Robert Baranyi, for which he was given the Nobel Prize in 1914.[3] He was a young doctor specializing in ear problems in the Clinic of Privy Councilor Professor Politzer, Vienna, and noticed that some patients became dizzy when their ears were flushed with water. In a search for clues for this phenomenon, he examined their eyes and noticed a jerky, rolling, motion called nystagmus.

One day a patient remarked that he only became dizzy when the water being used to flush his ears was not warm enough. He maintained that when he flushed out his own ears with warm water he never got dizzy. Following this lead, Baranyi washed out the man's ears with water that was too hot. Not surprisingly, the patient complained. Baranyi then observed the patient's eyes after flushing his ears with either hot or cold water, and noticed that the nystagmus which ensued was in opposite directions. Further work showed that even healthy people's eyes would rotate if their ears were washed out with cold or hot water. A consideration of all the results led to a recognition that there is such a thing as a middle ear.

Vision Processes

DAVID HUBEL TORSTON WIESEL

David Hubel and Torston Wiesel shared half of the Nobel Prize in 1981 for their discovery concerning the process of vision. Roger Sperry was given the other half. Because of a fortuitous event, Hubel and Wiesel were able to make a revolutionary breakthrough in our understanding of the subject.

Hubel had devised a microelectrode for detecting the response of individual cells in the cerebral cortex to various stimuli. Cats were the test subjects, and the microelectrode responses were monitored with an oscilloscope. With animals that were awake, Hubel succeeded in activating many cells by focusing the eye on moving spots on a screen. Some cells were selective in responding to movement of a spot in one direction across the screen, but not when it moved in the opposite direction. Many cells could not be influenced at all. Kuffler at Johns Hopkins had data indicating that the receptive field of each type of a certain cell was made of two mutually exclusive regions, a center and a surround, one excitatory and the other inhibitory. To impose a central figure surrounded by a clear area onto individual cells on the cat's retina, a black dot was glued onto a slide, which could then be inserted into a projector (technically, an ophthalmoscope). For the test that followed, the cat was anaesthetized. With the aid of an ophthalmoscope, the investigator could focus the spot, surrounded by a clear zone, on the desired part of the eye. This procedure worked well with the retina because those cells respond well to dots of light, but for cortical recording it was unsuitable. Not much progress had been made even after months of experimentation.

One day they made an especially stable recording in which the cell in question lasted nine hours, during which time they had a very different feeling about what the cortex might be doing. Here is Hubel's description:

> For three or four hours we got absolutely nowhere. Then gradually we began to elicit some vague and inconsistent responses by stimulating somewhere in the mid-periphery of the retina. We were inserting the glass slide with its black spot into the slot of the ophthalmoscope, when suddenly over the audio monitor the cell went off like a machine gun. After some fussing and fiddling we found out what was happening. The response had nothing to do with the black dot. As the glass slide was inserted, its edge was casting onto the retina a faint but dark shadow, *a straight dark line on a light background.* That was what the cell wanted, and it wanted it, moreover, in just one narrow range of orientations.

So their good fortune had several dimensions. Having the slide stick in the ophthalmoscope was essential to the finding. Equally necessary, however, was the orientation of the slide. The consequence was that a new concept of the vision process was available, and with it the Nobel Prize.

Nerve Growth Factor

RITA LEVI-MONTALCINI
STANLEY COHEN

For persistence and hard work, no better example could be cited than that of Rita Levi-Montalcini. Displaced from her home by the Fascist regime in Italy, she continued her research under primitive conditions in a makeshift kitchen laboratory in a farmhouse. Ultimately she came to the United States and established herself in Washington University in St. Louis. In 1986 she received the Nobel Prize in medicine or physiology, with Stanley Cohen, for the discovery of the Nerve Growth Factor. In that effort was one fortunate experience that could not have been predicted. The following narrative is based on her autobiography.

The human brain is said to contain a hundred billion nerve cells, strands of fibers connecting them, and billions of glial cells (supporting tissue) interposed among them. Early in this century Camille Golgi discovered a method of staining with a chrome-silver preparation to make nerve cells stand out in the smallest detail. Determining the interconnections among these myriad cells in an adult brain would be impossible. A chicken embryo, however, has only a few hundred cells, and the neurons in it are identical to what exist in all mammals, including man. Therefore chick embryos are used to study the development of the brain.

Levi-Montalcini's research involved the regulatory mechanisms governing the development of motor and sensory nerve cells in chick embryos. Her collaborator, Viktor Hamburger, had received a letter from a former student who described his work in grafting a mouse tumor onto the body of a chick embryo; he had noticed that in time the nerve fibers stemming from nearby sensory ganglia had merged into the mass of tumor cells. According to the student's interpretation, the tumor's properties and rapid proliferation produced conditions favorable to the fibers' growth, which in turn caused a greater number of nerve cells in the ganglia to differentiate. Levi and Hamburger decided to repeat the student's work. They took fragments of the tumors the size of a chick-embryo limb bud and grafted them onto the sides of three-day-old embryos. The result was what Levi termed an "extraordinary spectacle" (i.e., the mass of tumor cells was penetrated by bundles of brown-blackish nerve fibers from all sides). Furthermore, in no case had they established any connection with the cells. The effect was totally different than what one got when grafting a limb bud onto the embryo. A series of experiments followed, establishing the role of the mouse tumor in producing a halo effect on the fibers.

With the collaboration of Stanley Cohen, a young postdoctoral student, the product of the tumor was found to be a nucleoprotein. To study this nucleoprotein they needed to produce it in some quantity. The transplanting of mouse tumors had to be repeated many times to produce enough of the compound, because the yield was so small.

Cohen sensed the possibility that the nucleic component might be only a contaminant. Arthur Kornberg, a Nobelist, suggested that he treat the fraction with snake venom because it contained phosphodiesterase, which breaks down nucleic acids. Subsequent experiments with the venom-treated fraction showed an enormous increase in the activity of the tumor's Nerve Growth Factor (NGF). Later, assays showed that the venom's NGF was three thousand times stronger than that of the mouse tumor, which had prompted the whole study. Through some brilliant deductions the investigators were led to the mouse salivary gland where more NGF was found. Nowhere in Levi-Montalcini's account is the word "luck" used and, in fact, it would be out of place. But the unusually high concentration of NGF in snake venom had not been expected, and it was an important cog in their research.

Vitamin B_1

CHRISTIAAN EIJKMAN

Beriberi is a disease marked by inflammatory or degenerative changes of the nerves, the digestive system, and the heart. In most areas of the world it has probably been forgotten, but there was a time when it commanded attention. The term "disease" almost automatically suggests a bacterium or a virus as the culprit, and that was the case with beriberi, but fortuitous events showed that it was caused by a deficiency.

Christiaan Eijkman obtained his medical degree in 1883 and immediately served as medical officer to the Dutch East Indies, working in Java and Sumatra for two years. An attack of malaria forced him to repatriate to the Netherlands, where several years later he decided to train himself in bacteriology, which was an emerging science.[4] He went to work with Robert Koch in Berlin, where he joined two young men who were destined to head for the East in October 1886, where they would attack the problem of beriberi. At that time, he considered beriberi to be an infectious disease, but he did not succeed in producing the disease in animals by inoculation with the micrococcus thought to be the cause.

He was working in a laboratory in Batavia (Jakarta) on the island of Java. Chickens were the animals selected for this study, and a disease similar to beriberi seemed to develop spontaneously in them. His laboratory was on a military base that provided certain support; every day a laboratory assistant went to the mess hall to pick up the cooked rice that was the constant diet for the test subjects. After several months the chickens developed an unsteady gait, their combs turned blue, and some of them died. Repeated attempts to associate the symptoms with a disease were unsuccessful. Eventually, however, the chickens were no longer sick—all became healthy again for no apparent reason. Eijkman had no explanation for the turnaround but began to examine every factor, and in his words, "a chance happening put me on the right track."

There had been a change in command at the military base, and when the new supervisor of the mess hall learned that some of his cooked rice was being fed daily to chickens, he stopped the practice. He was willing to provide the rice but not the labor to prepare it, and as a consequence the chickens were fed raw rice rather than cooked rice. A retrospective study showed that the return to health of the chickens coincided with their consumption of raw rice. Eijkman recognized the possibility of a deficiency syndrome and was able to show that something in the whole rice was responsible for the health of the chickens. When fed whole rice, or injected with its extracts, the chickens remained healthy. For some time, Eijkman continued to believe that some chemical agent was causing the polyneuritis symptoms. In 1901 however, Grijns advanced the thesis that beriberi was caused by a nutritional deficiency. This fit in with the results Eijkman had achieved, and eventually that line of work resulted in the discovery of thiamin, vitamin B_1. An announcement of the Nobel Prize was made in 1929, but he was too ill to attend the ceremony.

Boron Chemistry

HERBERT BROWN

A memorable human interest story concerns Nobelist Herbert Brown of Purdue University.[5] He and his future wife were graduate students in the mid-1930s, a time when finances were scarce for everyone, but particularly for those having to meet school expenses. Brown was slated to receive his doctorate, and of course his girlfriend wanted to present him with an appropriate gift. With only two dollars to spend, she selected a book devoted to the subject of boron. Boron was not headline news in the chemical world in 1936, but Herbert Brown devoted his life's work to the subject after that and won a Nobel Prize. One factor in his selection for that honor was his development of boron hydride as a catalytic reducing agent, and it was in this regard that luck played a significant role. To preserve a batch of boron hydride he had prepared, he placed it in acetone; upon retrieving it later, he found that the acetone had been reduced to isopropyl alcohol. Organic chemists have capitalized on the reducing properties of boron hydride ever since. If he had chosen initially to store the boron hydride in isopropyl alcohol, there would have been no reduction and, maybe, no Nobel Prize.

REFERENCES

1. Blumberg, B. S. 1977. Australia antigen and the biology of hepatitis B. *Science.* 197: 17–25.

2. *The Nobel Prize Winners I, 1901–1944*. Pasadena, Calif.: Salem Press. 1991.

3. Stephenson, Henry Schuman. *Nobel Prize Winners in Medicine and Physiology, 1901–1950.* New York: Henry Schuman.

4. *Dictionary of Scientific Biography.* New York: Scribner.

Chapter 11

Technical Literature: Pluses and Minuses

"My mother made me a scientist without ever intending it. Every other Jewish mother in Brooklyn would ask her child after school, 'Did you learn anything today?' She always asked me a different question, 'Izzy, did you ask a good question today?'"

—Isadore Rabi

Scientists ask questions and report the results of their studies in the literature. One would hope that the literature reflects total objectivity in what is reported, but scientists are human beings subject to flaws in judgment. This brings to mind the pronouncement made by Bob Orben, a speech writer for President Gerald Ford: "Smart is when you believe only half of what you read. Brilliant is when you know which half."

Surprises and Limitations

Certain illustrious scientists did not regard a literature search as a necessary prelude to a new study. For example, Steve Brodie, who had many claims to fame, including being the mentor of Nobelist Julius Axelrod, had this to say: "Go to work in the laboratory; don't go to the library and read other people's frozen concepts."

There are problems other than frozen concepts associated with the literature. Because of space restrictions imposed on authors by the editors of journals today, what is in print might actually be misleading. Consider a hypothetical case. In the old German literature, you might find an account of the synthesis of a new compound in which the author would state: "I refluxed the mixture while I went out for a beer." It could be a short time or a long time, and that is inaccurate reporting. On the other hand, the reader understands the context in which the account is written. But, if the report stated that the refluxing was "for an hour," the reader might interpret that as being the optimal time. The point is that the clarity of a report may be decreased by the general practice of eliminating informality.

An associated problem is the lack of candor in report writing, which tends to foster boredom on the part of the reader. Also it can be misleading, as reflected in a statement by Arne Tiselius:

> A scientific paper … is usually of rather limited value if we want to know how things really happened. The facts and the conclusions may be presented in a perfectly logical order and with admirable elegance, but did it really happen in this way? The author writes his paper when he has arrived at the final results and then first realizes which way he should have gone. Probably the way he took was much more complicated, involving sidesteps, mistakes, disappointments, wishful thinking, and still more of that kind. Thus, a scientific paper mostly involves a certain kind of after-rationalization, naturally perfectly understandable and permissible. Scientific work is objective and its results should be devoid of the personality of the author. In a way this is a pity, since I believe that in science the human subjective factor is involved in the act of creation almost as much as in art, literature, and music.

Completeness

JAMES L. COX

The comments above touch on completeness, a theme developed here. An example is contained in one of the *Citation Classics* in *Current Contents,*[1] written by James L. Cox who had been a student at the University of Tennessee Medical

School in Memphis. In the spring of 1965, he was given the opportunity to work in a research laboratory with two cardiac electrophysiologists. Their intent was to demonstrate by chemical techniques that the injury to tissues in the heart following an attack was nonuniform. Cox recalled that he worked with a copy of a histochemistry textbook at hand to study these tissues. For a full year he had no success, until one day by mistake he adjusted the pH of the tissue-incubating solution incorrectly and as a result was able to demonstrate for the first time the nonuniformity of injury in the tissue. Subsequently he and a medical intern were able to reconstruct three-dimensional maps of myocardial infarctions that clearly documented the presence of viable zones that had been rendered anemic by insufficient blood flow. Their paper was published, but how informative was it? I looked up the original paper *and found no mention of the absolute importance of pH.*[2] The account did state that a buffer had been used, but there was no indication of the necessity of close control of the pH. So I wrote to Cox (at the Washington University School of Medicine in St. Louis) and asked whether there had been any circumstances that led to this significant omission. His reply was both interesting and rewarding. The article in question was the first that he had written, and it was just an oversight on his part that he left out the pH role. He went on to offer the following:

> Since you are writing a book on the influence of serendipity on scientific discovery, I cannot close without being sure that you know the story of Evarts A. Graham and his development of the oral cholecystogram test. If you notice on the letterhead, I am the Evarts A. Graham Professor of Surgery, this being Dr. Graham's institution during the first half of this century. Dr. Graham is generally considered to be one of the two or three most dominant figures in American surgery. He and one of his students, Dr. Warren H. Cole (who later became professor and chairman of the University of Illinois), were convinced that the oral ingestion of iodinated dyes would be concentrated by the gallbladder, and thus should outline the gallbladder if a standard abdominal X-ray were taken several hours after oral ingestion. They performed that study at this institution perhaps sixty years ago in dogs and were uniformly unsuccessful on innumerable occasions. One day they came into the lab, took an X-ray, and there was the beautiful gallbladder in a small dog! They immediately went back to the caretaker to see if anything had been done differently in that particular dog and his famous comment was 'nothing different, only that he did not get anything to eat last night.' Graham and Cole immediately realized that in order for this test to work, the subject had to be kept NPO and thereafter the 'Graham-Cole Test' for the visualization of gallbladders became a standard for the next half century.

It is worth noting here that three other discoveries mentioned in this book depended on the roles played by janitors or caretakers (see accounts of Richet, Eijkman, and Nalbandov).

Keeping up with the literature is not easy. Fergus Campbell, at the Physiological Laboratory at Cambridge, asked Edgar Adrian how he dealt with the exponentially increasing research literature, and his reply is worth thinking about: "It is only necessary to read the first paper which proposes the problem and the last paper that solves it. The ones in between do not matter."

Ignorance as a Potentially Positive Factor

LUIS ALVAREZ
T. MICHAEL DEXTER

Nobel Prize winner Luis Alvarez pointed out a situation in which ignorance of the literature was beneficial when he was developing an understanding of how a certain clay layer might have originated. His son, Walt, had found this layer in a rock, between a bottom layer of white limestone filled with foraminifera and an upper layer of red limestone. They were searching for metals of the platinum family for use in trace element analysis. Such metals occur relatively abundantly in meteorites. They found, however, that iridium was a better choice than platinum. Iridium has an enormous cross section for neutron capture. This work on iridium led to the awareness of its occurrence in the K/T boundary marking the Cretaceous-Tertiary period. Two men at the University of Chicago had independently attempted a tracer method, similar to that used by Alvarez, as a way of measuring sedimentation rates, but had not been successful. Alvarez made the point that if he had known of their work, he might not have followed his instincts, which turned out to be successful.

There was no question in the mind of T. Michael Dexter about the potential merit of ignoring the literature. An article of which he was the principal author was entitled "Conditions Controlling the Proliferation of Haemopoietic Cells in Vitro."[3] It turned out to be a *Citation Classic:*

> Fortunately I did not know that many such attempts had been made in the past, all unsuccessful. A thorough literature search might have discouraged me from undertaking the task, but computers were not accessible then.

Problems with Peer Review

No one could argue with the basic premise supporting peer review. Problems arise, however, because reviewers are fallible human beings who may be deficient in objectivity. Luis Alvarez, in his entertaining book *Adventures of a Physicist,* made this criticism:

> In my considered opinion the peer review system, in which proposals rather than proposers are reviewed, is the greatest disaster to be visited upon the scientific community in this century. Perhaps the most serious defect is that it effectively prevents scientists from changing fields. So I see lots of my friends spending their whole scientific lives essentially repeating their PhD thesis.

One reason for this is that officials in agencies that offer grants tend to prefer extensions of existing programs rather than untried innovative approaches. An encouraging sign opposing this situation was a letter sent to *Chemical and Engineering News* by a staffer at the U.S. National Science Foundation (NSF) in Washington, DC. It stated that research proposals of a more novel nature would be welcomed by that agency. The letter went on to say that despite this known position of NSF, few of the research proposals they were receiving reflected that stance. I called the letter-writer to compliment him on what appeared to be a hopeful portent. Several years later, when I chanced to call him again on another matter, he had returned to academic life. Recalling that I had spoken to him when he was at NSF, I asked whether his colleagues there had endorsed his suggestion that they be more receptive to new ideas. Unfortunately, they had not. In his judgment, at least, the turn-the-crank approach to funding research proposals at NSF still favored continuances of ongoing programs.

Stifling initiatives is not the only problem with the peer review system because, in many cases, it has not worked well. Searching through hundreds of *Citation Classics,* I was impressed by the significant number of cases in which these often-referenced authors mentioned the difficulty they had in getting their work published. A *Citation Classic,* by definition, is an important paper, but in a study made by the publisher of these articles, six percent of those authors mentioned having difficulty with the reviewers (my guess would have been a higher percentage, but I was not interested in these statistics at the time of the study). The following are samples of unfortunate peer reviews.

John W. Kebabian and Donald B. Calne wrote a review on dopamine that was published in *Nature* in 1979 after being rejected by *Science.*[4] The rejection letter stated that "it was not of sufficient general interest" and "it should not be published in any journal." In view of the fact that it was subsequently referenced more than two thousand times, many scientists felt otherwise.

A landmark article by Lineweaver and Burk on enzyme kinetics appeared in *Journal of the American Chemical Society.*[5] But, its road to publication was not smooth. The article was rejected by all three reviewers to whom it was sent initially. After an appeal by the authors, it was again rejected by three supposedly different reviewers. Eventually the editor of *JACS* disregarded the objections and published it. It has been referenced thousands of times.

In 1975 an article published in *Lancet* by two brothers named Miller concerned the role of high density lipoprotein (HDL) in the blood. Reviewers had rejected it, but the editor disagreed with their decision and found it to be worthy of publication. In a count made several years ago, it had been referenced over five hundred times.

Guido Maino suffered the indignity of rejection only one time in his career. The *Journal of Experimental Medicine* turned down Majno's article concerning the role played by histamine in the permeability of small blood vessels. The article was accepted later by the *Journal of Biophysical and Biochemical Cytology* and was subsequently referenced more than six hundred times.[6]

An interesting commentary on the review process is provided by J. W. Paulley, who coauthored an article with J. P. Hughes on giant cell arteritis.[7] This work was published by the *British Medical Journal* and has been referenced more than two hundred times. Here is part of Paulley's comments written in 1987 on the article published in 1960:

> In a lighter vein, it may be a comfort for frustrated authors to know that the article was rejected by one journal and then by the one that eventually accepted it. Fortunately for me, editors were then more independent of their expert assessors than they are today. I pleaded to my editor that the fact that a cardiological advisor did not like my inclusion of aortic and coronary artery involvement in GC(T)A because he had not seen it, or bothered to read about it, was insufficient reason for rejection. I also reminded him that the speed of a convoy was that of the slowest ship and that if assessors' entrenched opinions were always to prevail, much deserving work would be lost.

An obviously disbelieving reviewer for the journal *Diseases of the Chest* was responsible for the rejection of an article by Leo Schemroth and colleagues. Entitled "Immediate Effects of Intravenous Verapamil in Cardiac Arrythmia," the paper was rejected on the grounds that: "1) There is no such thing as a Ca-blocking agent, and 2) digitalis acts in precisely the same manner." Schemroth pointed out in his short narrative that all twenty of the original subjects were fully "digitalized" and the effect noted was clearly over and above that of digitalis. Subsequently the paper was accepted by *Cardiovascular Research,* and at the time of its listing as a *Citation Classic* had been referenced more than three hundred times.[8]

If ever anyone had problems with the peer review system, it was Hannes Alfven. He was a man far ahead of his time and consequently was criticized widely by his contemporaries in the fields of physics, astrophysics, and space sciences. Because of this, he had difficulty in getting his writings published in the more reputable journals. He was vindicated by being awarded the Nobel Prize in physics in 1970. An interesting summary of his life was written in 1988 (by Anthony L. Perratt, "The World and I"). According to Perratt's account, Alfven said, "I have no trouble publishing in Soviet astrophysical journals, but my work is unacceptable to the American astrophysical journals." Alfven made an important point about the peer review system with this statement: "The peer review system is satisfactory during quiescent times, but not during a revolution in a discipline such as astrophysics, when the establishment seeks to preserve the status quo."

My favorite on the subject of reviewer fallibility concerns a report by Rohrer and Binnig. Here is part of the reviewer's venomous appraisal:

> The paper is virtually devoid of conceptual discussion let alone conceptual novelty ... I am interested in the behavior of the surface structure of gold and the other metals in this paper. Why should I be excited about the results of this paper?

It happened to be a paper concerning the scanning tunneling microscope and, you may recall, the Nobel Prize was awarded to Rohrer and Binnig for this great advance.

These are enough examples to illustrate that reviewers can make mistakes. To the extent that one might equate reviewers with experts, a comment by Roger Altounyan is worth mentioning: "Beware of the expert; by the time he is generally regarded as such he should usually be referred to in the past tense."

Pirates and Ignorance

Associated with the peer review process is the possibility that unscrupulous reviewers pirate ideas from the articles submitted to them without giving credit to the author. Several years ago Rustum Roy of Penn State University chose to hold a press conference to announce a recent discovery of a procedure for making synthetic diamonds. Roy had balked at having an article pass through peer review lest the findings may be used by competitors who may be serving as evaluators.

John Christman used to tell of his complaint with a reviewer who criticized his inclusion of a personal recollection in a report he had written. The subject concerned the preparation of a biological stain that Chris had prepared. The stain was divided into several lots, and the personal observation was that the batch left near the sunlight coming through a window had worked best. The reviewer saw no merit in what he termed "your meteorological observation," being blithely oblivious to the possibility that the action of the sun's rays was beneficial to the performance of the stain.

A Salutary Case Regarding the Literature

Not all the news about publication practices is bad. The following story involves an instance of outstanding collaboration between two research groups:

Joshua Lederberg of Rockefeller University led a study that successfully uncovered a mechanism for selecting organisms that grow poorly compared to the wild type. The converse, that of selecting conditions to favor a fast-growing organism, is generally much easier. When Salvador Luria, then at the University of Illinois, visited Lederberg's laboratory he pointed out to Lederberg and his collaborator, Norman Zinder, that Bernard D. Davis had essentially made the same discovery at Harvard Medical School. The basis for both studies had been that penicillin kills only rapidly growing cells, leaving stationary cells alive. Rather than race for publication before Davis did, Lederberg and Zinder contacted Davis and suggested that they publish jointly, with both reports being submitted to the *Journal of Biological Chemistry.* They were rebuffed by a reviewer who maintained that the reports had too little "chemistry" in them. Fortunately, the editor of the *Journal of the American Chemical Society* took a broader view and accepted the articles for publication.[9] Lederberg suggested that both papers be printed side by side so that reprints of both could be readily available to interested parties. This spirit of cooperation is not always evident in science (you may recall the constant rebuffs suffered by Eugene Roberts when he demonstrated the role of GABA in the brain).

An Interesting Case History

Research involving freeze-drying demonstrates the value of ignoring—and of consulting—the technical literature before starting an experiment. It was at the Mill Hill Laboratory outside of London in 1949 when researchers Audrey Smith and C. Polge were interested in preserving the sperm of fowl. They had decided to do this by freeze-drying, a procedure that was not brand new, but less known than today.

As a prelude to this approach they analyzed the sperm and found that it contained a significant quantity of fructose. Having learned this, they decided to prepare various solutions of fructose as potential freeze-drying agents, which were stored in a cold room. This room was shared with another laboratory several miles away. Bottle after bottle of fructose solutions was tried with no success. Then one bottle that had no label was the one that worked. Upon further examination it was found that the bottle did not contain a fructose solution, but instead contained glycerin. Furthermore, it wasn't even their bottle; it belonged to the laboratory several miles away. It's particularly interesting to note that it was the custom at Mill Hill to *not* consult the literature before beginning experimentation according to Jim Lovelock.[10] It is fortunate that was the case for, prior to 1949, there were literature references to the failures in freeze-drying of tissues. Had the Mill Hill researchers known of these failures they may never have begun with that approach.

Years later, when the medical profession was making history with organ transplants, it was clear that a method for preserving corneas was needed if they were to become useful as transplants. According to Jim Lovelock's recollection (as stated in a personal letter to me), a researcher named Elder recalled the success with freeze-drying tissues at Mill Hill and suggested that freeze-drying with glycerin be tried as a way to preserve corneas. It worked, and it is my understanding that freeze-drying is still used to preserve corneas.

So there are now people who have the joy of being able to see their loved ones, to read a book, or drive a car, all because of the accident of a label falling off a bottle at Mill Hill. That progression, in turn, was possible because one group made a practice of reading the research literature and another one didn't. Such are the ways of science.

REFERENCES

1. Cox, James L. 1984. *Current Contents (CP) No. 16.* April 16, pg 16, referring to Cox, J.L. et al. 1968. The ischemic zone surrounding acute myocardial infarctions. *Amer. Heart J.* 76: 650–659

2. Cox, James L. et al. 1968. The ischemic zone surrounding acute myocardial infarctions. *Amer. Heart J.* 76: 650–659.

3. Dexter, T. Michael. 1987. *Current Contents.* May 2, 1988, pg 18, referring to Dexter, T.M. et al. 1977. Conditions controlling the proliferation of haemopoietic stem cells in vitro. *J. Cell. Physiol.* 91: 335–944.

4. Kebabian, John W. and Calne, D. B. 1979. Multiple receptors for dopamine. *Nature.* 277: 93–96.

5. Lineweaver, Hans. 1984. *Current Contents* (LS). March 18, 1985, pg 19, referring to Lineweaver, H. and Burk, D. 1934. The determination of enzyme dissociation constants. *J. Amer. Chem. Soc.* 56: 658–666.

6. Maino, Guido. *Current Contents.* (Citation unavailable), referring to Maino, G. 1961. Studies on inflammation: 1. The effect of histamine and serotonin on vascular permeability: An electron microscopic study. *J. Cell. Biol.* 11:571–605.

7. Paulley, J. W. 1987. *Current Contents* (CM). Dec. 7, 1987, pg 12, referring to Paulley, J.W. and Hughes, J. P. 1960. Giant cell arteritis, or arteritis of the aged. *Brit. Med. Jour.* 5212, 1562–1567.

8. Schemroth, Leo. et al. 1972. Immediate effects of intravenous verapamil in cardiac arrhythmia. *Brit. Med. Jour.* 1 (801), 660.

9. Lederberg, J. 1987. *Current Contents.* Aug. 17, 1987, pg 16, referring to Lederberg, J. and Zinder, N. 1948. Concentration of biochemical mutants of bacteria with penicillin. *J. Amer. Chem. Soc.* 70: 4267–4268, published simultaneously with Davis, B. D. 1948. Isolation of biochemically deficient mutants of bacteria by prenicillin. *J. Amer. Chem. Soc.* 70: 4267–4268.

10. Lovelock, James. Personal communication.

CHAPTER 12

CHANCE AND FINDING ONE'S NICHE

"Science is the endless quest to look behind the walls of superstition and ignorance to understand our world and those who inhabit it, where we came from and where we are going. Science is indeed magnificent entertainment! Why hadn't my teachers ever told me about this?"

—Duncan Blanchard

Science can be magnificent entertainment, indeed. Developing an understanding of a process, regardless of whether or not others consider it important, brings great satisfaction. If that understanding improves human health or generates financial riches, then the satisfaction is so much greater. Throughout the history of science, various factors have prompted certain people to become scientists. In light of this book's focus on serendipity, it seems proper to explore whether chance might have been involved in the career decisions of a number of renowned scientists.

The Role of a Mother in a Career Choice

M. Gordon Wolman, chair of the Department of Geography and Environmental Engineering at Johns Hopkins University (Baltimore, Maryland) said that he didn't know whether chance actually directed him into his field. He did have this to say, however:

> Chance combined with family interests placed a city boy (me) on a farm because my mother thought that I should know that milk did not come out of a bottle. I took to it. This experience was reinforced by my father's professional involvement in the field of resources and specifically water resources. I chose geology of the surface of the earth for graduate school for my direction. That was planned.

But then he went on to say that it was chance that put him into contact with Luna B. Leopold, a hydrologist for whom he went to work when he was in graduate school.

A Case where Familiarity Bred Content

Alfred G. Knudson of the Fox Chase Cancer Center in Philadelphia, Pennsylvania, had intended to work for a PhD in biology. He had never intended to go to medical school, but World War II came along and prevented him from pursuing the career he had chosen. Instead, he was advised by the Navy to go to medical school in its V-12 program. He liked medicine and decided to make a career of it. Regarding whether chance played a role in his research findings he had this to say:

> My chief work has concerned a theory about the genetic basis of cancer, a subject I had pondered for some time, and specifically considered retinoblastoma, a tumor of children. Chance entered the picture when I moved to an institution that, unknown to me, had collected the data I needed to test my underlying hypothesis. As far as I can recollect, chance did not play a role in generating the hypothesis.

The Importance of a Girlfriend's Recommendation

George N. Somero, professor of biology and the department chair of marine biology at the University of California, San Diego, California, wrote a long letter with details of how chance affected both his choice of work and his research results. When he was at Stanford University, his girlfriend asked him about his postdoctoral plans, which were "fuzzy" at that time. She suggested that he contact a friend of hers whom she had known at Duke University (Durham, North Carolina), Dr. Peter Hochachka. Hochachka became Somero's postdoctoral advisor, major mentor, and lifelong friend. Subsequently Somero and Hochachka published two books that have helped to shape their field of "comparative or adaptational biochemistry." As for results based on any serendipitous circumstances, here are his words:

> Since coming to Scripps Institution (La Jolla, California) unexpected good things have continued to happen in my career. Having a faculty position at the world's best oceanographic institution probably makes the word 'serendipity' questionable in the sense that the support offered by Scripps (excellent labs, good colleagues, superb students, and access to virtually any environment and organisms in the ocean) can't help but make good things happen. The closest thing to a serendipitous event during my life at Scripps happened 2500 meters below the sea, near the Galapagos Islands. This was the unexpected discovery of the unusual animals living at the deep-sea hot springs. This 1977 discovery led me to submit an NSF (National Science Foundation) proposal for work with these animals. We brought back to Scripps for biochemical analyses some of the first specimens of the large vent tube worms and clams. At about the same moment that the frozen specimens arrived from Galapagos, Dr. Horst Felbeck arrived from Germany to work as a post-doc in my lab. When we discussed research projects, I recommended that he examine the metabolic properties (by doing enzymatic analyses of the frozen specimens) of these vent animals. Horst discovered the symbiosis between these animals and sulfide-burning bacteria. His discovery in my lab—which, of course, led to some rubbing-off of his fame on me (Merton's "Matthew Effect in Science")—has led to a very active and productive research program on the vent creatures. Eleven years ago, we didn't know they existed. Now there are dozens of marine scientists, including myself, actively working on these organisms.

An Astronomer Finds His Way

At the other extreme of earth's boundaries is the case of an astronomer. Alexander Dalgarno, the Phillips Professor of Astronomy at the Harvard-Smithsonian Center for Astrophysics (Cambridge, Massachusetts) began his career as a mathematician. As a graduate student he came to work in atomic and molecular physics simply because Sir Harrie Massey offered him financial support through a fellowship. He knew nothing about the subject at the time. His first faculty position was with David Bates, who was a giant in the field of atmospheric science. It was through interacting with Bates that Dalgarno became involved in studies of the upper atmosphere, and that led to the still more exotic environments found in astronomy.

A Seminar Turned Out to Be Important

Brigitte A. Askonas, head of the Division of Immunology at the National Institute for Medical Research in London, recounts an unplanned circumstance that led her to immunology:

> I had been involved in the synthesis of milk protein with Dr. Thomas Work (using radiolabeled amino acids) and it was quite difficult to isolate the beta-lactoglobulin in numerous very small samples. I then heard a seminar by Dr. Humphrey here at Mill Hill. In those days in the early 1950s it was usual to give a demonstration during a seminar. Dr. Humphrey showed an antibody/antigen precipitate by mixing a soluble solution of pneumococcal polysaccharide with the clear serum of a rabbit hyper-immunized with pneumococci. There was an immediate precipitate. This demonstration prompted me to decide to shift my studies to antibody synthesis rather than milk protein synthesis. All I had to do was to immunize animals with an antigen and then look at antibody formation in cell suspensions by purifying the antibodies after addition of the antigen.

She worked on antibody synthesis for many years and has remained in the study of immunology for fifty years.

The Importance of a Teacher

Allan S. Hay, chair of polymer chemistry at McGill University (Toronto, Ontario, Canada) credits one of his professors with giving him the desire to work in organic chemistry. While a student at the University of Alberta (Edmonton, Alberta, Canada) he took an organic chemistry course under Reuben Benjamin Sandin and found it so interesting that he devoted his life to the subject. Following his postgraduate work at the University of Illinois, Hay joined General Electric at its research laboratory in Schenectady, New York, because he was told that he could work on anything he pleased! Hay chose to study the catalytic oxidation of organic compounds with oxygen, and in the course of those studies synthesized a polymer (polyphenylene oxide) that became a large financial success because of the diverse uses to which it could be applied.

Circumstances Define the Moment

Robin Holliday, Senior Research Fellow at the CSIRO (Commonwealth Science and Industrial Research Organization) Division of Molecular Biology (Sydney, Australia), fortuitously arrived at Cambridge University in 1952 just as Watson and Crick were unraveling the structure of DNA. By 1954 his choice for a career in genetics was secured. It was he who discovered the process by which DNA molecules combine, and how this process makes each one of us a unique individual. It has been said that practically the whole biotechnology industry has been based upon Holliday's findings.

With Medicine Not an Option, Why Not Work on Plants

Donald Walker retired in 1989 as professor of biogeography and head of the Department of Biogeography and Geomorphology in the Research School of Pacific Studies at the Australian National University (Canberra, Australia). Until he became established in his field at about the age of thirty, his career was directed more by opportunity than by plan. During World War II many outstanding women teachers were brought to his school to replace the men called into service. Among these was a biologist (her name was not given) who was the wife of one of Britain's leading plant ecologists. It was she who nurtured his interest in natural botany. Walker had planned to enter the university to become a veterinary surgeon, but because of the limited number of vacant slots, that was not possible; veterans returning from World War II had priorities. This forced Walker to be content with an education as a biologist.

The military played a further role in his career when he had to go on active duty. During his two years of service, the army discovered a need for an ecologist in the Malayan war. Walker was given that assignment, which was what led to his expertise with tropical plants. Regarding the role of chance in his research, he cites the restoration of diplomatic relations between Australia and China in the early 1970s as the opportunity to test an idea that could not have been tested otherwise.

In his fascinating and lengthy reply to my letter, he stressed the merit of remaining receptive to the unexpected event, and to avoid becoming too specialized. That theme can be found in the following narrative also.

A Recruit for Linguistics

Morris Halle, Department of Linguistics and Philosophy at Massachusetts Institute of Technology (Cambridge, Massachusetts), credits an unexpectedly early discharge from the service with his decision to enroll at the University of Chicago in the spring of 1946. He had not considered Chicago before, but its attraction was a trimester system starting in April. To enter most other colleges, Halle would have had to wait until September. There at Chicago, a second chance event was that he met a visiting linguist who urged him to go to Columbia to study with Roman Jakobson. That he did. When Jakobson left Columbia to go to Harvard, Halle went along. While at Harvard he got, from a third professor, a totally new concept of what linguistics should achieve, and it was this concept that set Halle on a course for life.

An Effective Teacher Leads to a Career in Genetics

O. H. Frankel, of the aforementioned CSIRO in its Division of Plant Industry, decided to study agriculture on the basis of a simple observation at the age of eighteen; there was hunger in the world which must be satisfied. At that time he knew nothing about agriculture, but he left his native Vienna to attend the Agricultural University of Berlin. To use his words:

> Attendance at lectures not being compulsory and most professors having written books on their subject, I decided to skip the former and study the latter, except for a few lectures to get to know them. This sealed my fate to genetics. The professor of genetics, Erwin Baur, was a fascinating lecturer, and the subject—I had not even heard of it before—quite captivating. I went to every lecture, studied his book, and he accepted me for a doctor's thesis before having finished my first degree. What would I have become if he had been dull?

Regarding the effect of serendipity on his work, Frankel mentioned that he went to New Zealand where during his first or second harvest of wheat his only assistant was a boy about fourteen years old. The youngster proved to be a real help, however, when he drew Frankel's attention to a head of wheat that had striking characteristics that identified it as a "speltoid." More interestingly, it was partially sterile in a definite pattern. This led to a long research effort on the genetics and physiology of floral development in wheat, the last paper having been published in 1975.

Opportunity for Study Leads to Nobel Prize

It was the option to work toward an advanced degree while being employed that led Herbert Hauptman to the Naval Research Laboratory (Washington, DC) in 1946. Hauptman had his master of arts degree in mathematics when he was discharged from active duty from the Navy and was committed to getting his PhD. NRL's policy of encouraging advanced study allowed him to work part time while he attended the University of Maryland, where he received his doctorate. Hauptman wasn't fussy about his work assignment; any field would do. Chance would play a significant role in his entry into the field of X-ray diffraction. He started in the Electron Optics Branch under Herbert Friedman, known internationally for his work in upper atmosphere research, and a member of the National Academy of Sciences. Ultimately, Hauptman teamed up with Jerome Karle in a study of electron diffraction, which led to a Nobel Prize for them. An interesting sidelight of that prize was that Karle learned of it during an international flight back to the United States. Somehow the airline learned of the announcement and the pilot of the plane announced to the passengers that a Nobel Prize winner was among them. The airline served champagne to all in the plane in tribute to its distinguished guest.

An NRL Policy Benefits Another

The NRL policy of encouraging graduate study led to the development of a valued researcher in the person of James Griffith. His father was employed by the United States Treasury Department, and the family had moved to a neighborhood "just over the hill" from NRL. Griffith, scheduled to begin graduate studies at the University of Maryland (College Park) in the fall of 1955, applied for summer employment at NRL. His plan had been to get his doctorate and then return to his native Alabama to work in rocketry at Huntsville. Being able to do his graduate studies at Maryland while working part time at NRL resulted ultimately in his spending thirty-six years there. He made significant contributions in the synthesis of improved linings for fuel tanks and for other special uses. Included in his research have been polyurethanes, latex materials, epoxies, and fluorinated resins. Applications of his research have been made in submarines, airplanes, and satellites. A coating he developed played a major role in the development of a stealth plane that did not reflect radar waves. In addition to many awards presented by the Navy, he won the prestigious Hillebrand Award presented each year by the Chemical Society of Washington.

A Career in the Study of Crystal Structure

George D. Watkins, of the Sherman Fairchild Center for Solid State Studies at Lehigh University (Bethlehem, Pennsylvania), was introduced to the field of defects in crystals by a serendipitous event. His intent was to devise a clever means of measuring nuclear quadrupole moments by applying uniaxial stress to an alkali halide crystal. To his surprise, the lines did not split, but rather irreversibly decreased in intensity. By the time his thesis was completed, he had learned about "dislocations," plastic flow, and work hardening, and had discovered a sophisticated and unique way to monitor the internal strains associated with dislocations in a crystal.

A Natural Talent Emerges

To those who are knowledgeable about air/sea interactions, the name Woodcock will be familiar. Alfred Woodcock dropped out of high school and drifted from one job to another before deciding to be a farmer. He took a two-year course in practical farming at the Massachusetts Agricultural College (now the University of Massachusetts) and was then hired by a farmer. This was no ordinary farmer but, rather, a graduate of Harvard who had ample means to own a farm and a yacht. In the late spring of 1930, Woodcock and others accompanied the farmer on a leisurely sail along the Massachusetts coast to Cape Cod, where they dropped anchor at the sleepy village of Woods Hole. Woodcock needed a haircut and entered the town's only barber shop for a visit that was to change his life dramatically.

While awaiting his turn he asked the man seated next to him about the identity of a large brick structure being built at the waterfront. It so happened that the seatmate was the captain of a U.S. fisheries vessel, and knew that the structure was to be an oceanographic institution. Its director was to be Henry Bigelow of Harvard, and at that time there was a need for crewmen to travel to Copenhagen to bring back the *Atlantis,* a new research vessel being built for the facility. The idea of extensive seagoing experiences appealed to Woodcock, and he approached a twenty-five-year-old protégé of Bigelow, named Columbus Iselin, for a job. Iselin offered him a job at forty-five dollars per month, feeling that he would be a steadying influence among what he feared was a turbulent crew. In the spring of 1931 Woodcock was among those who took control of *Atlantis,* the largest such vessel in the world, with a length of 142 feet. He was a keen observer and learned much about the sea on the trip back to Woods Hole. An expanded version of Woodcock's career and how it meshed with others whose interests included air/sea interactions is found in chapter 3.

Looking back on these early experiences, Woodcock said:

> I had little or no real inkling of the real meaning of science before I arrived in Woods Hole on the *Atlantis*. However, during this period I was learning that my interest in farming was really an interest in nature, whether airborne, waterborne, or ashore. From the beginning I did not like school, unfortunately. Why, I am not sure. For years after leaving school a recurring nightmare was to find myself back in the classroom again.

Disparate Forces Lead to Air/Sea Studies

A person whose professional life was closely intertwined with Woodcock's was attracted to a scientific career in an unusual fashion. Duncan Blanchard's early interest was in art. Upon graduation from high school, he became an apprentice machinist at the General Electric Company factory in Pittsfield, Massachusetts, where he assisted in building transformer cases for the war effort. Ultimately, with the Draft Board about to embrace him, he enlisted in the Navy's V-12 Officer Candidate Program, which was starting in July 1943. This program led to his entrance into Harvard, and then to Tufts University (Medford, Massachusetts), where his studies were principally in mechanical engineering and naval science, the latter becoming of great interest to him. By the time he was commissioned and sent to Guam, World War II was over and life there "was right out of Rodgers and Hammerstein's *South Pacific*," with trade winds swishing through the coconut palms every night and slot machines available at the Officers' Club overlooking the beach. Making his way one night along a path through dense tropical undergrowth to the Officers' Club, he saw something among the wet leaves lining the path—a very soggy book entitled *Great Men of Science* by Grove Wilson. After allowing it to dry sufficiently so that it could be handled, he started reading it the next night. For the first time he realized that science was not the embodiment of the boring experiments he had endured in the chemistry laboratory or the memorization of numerous equations. The quote at the beginning of this chapter expresses his fascination with the subject. His newfound interest in research played a key role in the field of air/sea interactions described in the next chapter. Blanchard is now retired from the Atmospheric Sciences Research Center at the State University of New York in Albany. At the urging of the American Meteorology Society, he wrote a book to increase an understanding of the atmospheric sciences; *From Raindrops to Volcanoes* is a gem to be enjoyed by anyone with an interest in science.

Safety Takes a Backseat in the Pursuit of Science

One of the more improbable entries into a scientific career was that of Joseph Kuc, Department of Plant Pathology, College of Agriculture, at the University of Kentucky (Louisville). Reminiscing about his career in an issue of *Current Contents,* Kuc said:

> My interest in plant disease resistance started forty-four years ago on the fire escape of an apartment house located on Jackson Avenue in the Bronx (New York). Much to the chagrin of the fire inspector, the fire escape had pole bean vines growing up the ladders, tomato vines tied to the railings, peppers, and an occasional zinnia growing in pots where space was available. My interest in chemistry and biology as related to plant disease resistance was nurtured at the Bronx High School of Science and clearly focused by working at the New York Botanical Gardens during the summer while studying for a BS in biochemistry at Purdue University (Lafayette, Indiana).

Had there been a fire at his apartment house, with complications arising from the intensive agriculture on the fire escape, the world may not have heard of Joseph Kuc.

From Music to Metabolism

As mentioned earlier, much of the material relating to the National Institutes of Health that is mentioned in this book has been taken from Robert Kanigel's excellent book *Apprentice to Genius.* In the previous chapter Solomon Snyder was mentioned, but how he became associated with NIH is an interesting story. According to Kanigel's book, Snyder was giving guitar lessons in a shop at 18th and M Sts. NW in Washington DC, approximately ten miles from NIH in Bethesda. One of those who came into the shop was Donald Brown, a medical doctor from the University of Chicago, who was to be a research associate at the National Institute of Mental Health (a part of NIH) for two years. Upon inquiry about the cost of the lessons, he decided that they were too high, so Snyder taught Brown on the side and ultimately his interest in medicine became apparent. Brown worked in the laboratory of Marion Kies and paved the way for Snyder to become a summer student there. In time, Snyder became associated with Julius Axelrod, who was later awarded the Nobel Prize.

An Influence from World War I

When Alan Lloyd Hodgkins, emeritus professor of biophysics in the University of Cambridge, was a student, he came across papers written by Keith Lucas, a close friend of his father. Both men had been killed in World War I. Because Lucas had done work with nerve cells, Hodgkins became interested in the subject and extended Lucas's work. Today Hodgkins is well-known for his study of nerve impulses, and he was named a Nobel Laureate in Physiology or Medicine in 1963.

Chance in the Choice of a Thesis Topic

Geoffrey Eglinton, professor of organic geochemistry, University of Bristol (Bristol, UK), related that chance definitely played a major role in his choice of a thesis topic. It depended largely on who was available as a mentor. He had finished his undergraduate degree at Manchester, and a new team of professors and lecturers had just arrived from London. He made a change from the synthesis of acetylene to molecular geochemistry, which dated back to an experience he had as an undergraduate.

While walking in the hills near Manchester, he encountered a deposit of bitumen at a place called Windy Knoll. This aroused his curiosity, for the substance was clearly organic in nature. He made a mental note that at some point he should try to determine its history.

About the same time he heard Melvin Calvin, and later Sir Robert Robinson, speak at Glasgow, and both of them indicated the great potential of organic compounds as sources of information in the geosphere. About 1960 he resolved to look at the occurrence of organic compounds in rocks, crude oils, etc. His work was furthered by several additional stimuli. Charles Chibnall, retired at Cambridge (where he was noted for his work in plant wax biochemistry), saw a paper coauthored by Eglinton and wrote to suggest that his collection of plant waxes would be useful with the new analytical techniques that Eglinton's group had described in 1960. Furthermore, the timely development of organic mass spectrometry at Glasgow by Roland Reed aided such analytical endeavors.

In the course of his education, Eglinton traveled to Ohio State University (Columbus) to spend time at Mel Newman's laboratory, where Eglinton developed his expertise in infrared spectra. He also worked with Melvin Calvin at the University of California (Berkeley) where he met another geochemist, L. R. Fyffe.

In Eglinton's career, it wasn't a matter of just one person influencing his career. There were random acquaintances made, but the whole course of events began with an astute observation of a deposit of bitumen at Windy Knoll.

A Switch to Virology

Sir Gustav Nossal, a Nobel Prize winner and director of the Walter and Eliza Hall Institute of Medical Research in Melbourne, Australia, cited two accidental events in determining the direction of his career. Nossal had intended to follow in the footsteps of an older brother, a biochemist, after finishing the medical course he was taking at the University of Sydney. The dean of the faculty of medicine insisted that it would be a waste of his medical training to go into pure biochemistry. Why not, the dean suggested fortuitously, link up with a virologist on the faculty who was doing microbiological work from a virology standpoint? Nossal took the advice, and in 1951 learned the fundamentals of virology. During that year he was introduced to Sir Macfarlane Burnet, a prominent Australian virologist and Nobel laureate, who impressed Nossal greatly with his research. After completing his residence training, Nossal moved to Burnet's institute, a move that altered his whole career by introducing him to the uniqueness of each single cell in immunology. Nossal described his experiences:

> Working with single cells has been technically difficult and has also allowed insights into the immune system that are subtly different from those that have been gained by more conventional tissue approaches. The way in which I got into single cells, however, in 1957, also owes much to two chance events. The first was the reading of a paper by Dr. Vogt dealing with virus replications inside single cells. The elegance of the method struck me as formidable. I quickly thought that looking at antibody formation at the single-cell level would be a clever way of approaching some of the controversial theories of the day, but who was going to help to teach me the elements of single cell micromanipulation?
>
> It just so happened that a brilliant bacterial geneticist, Joshua Lederberg, was visiting the Walter and Eliza Hill Institute in 1957 as a Fulbright Visiting Professor. He was an expert in micromanipulation. He had never thought of working in immunology before, but he was generous enough to be extremely interested in some of the experiments I was proposing and to spend many hours teaching me the fundamentals of micromanipulation. Were it not for that exposure, I might never have discovered that one cell always makes one antibody, and might not have pursued the pathway that, somewhat circuitously, led to the validation of the clonal selection theory, on the one hand, and to the discovery of monoclonal antibodies on the other.

A Tough Move, a Happy Find

Walter Bodmer, eminent geneticist at the Imperial Center Research Fund Laboratories in London, is aware that chance plays a large part in anyone's life and suggests that it is hard to know where it begins or ends. Because his father was Jewish, he had to leave Germany when Hitler came to power. Young Bodmer was brought up in England. Here is part of his letter to me:

> Was it chance that my parents had a friend who taught at the school I went to, who urged that I take up mathematics if I was good at it, which indeed I did, although my father, a doctor, would dearly have wanted me to study medicine? To this day I am not sure really why I did not. Having taken an interest in statistics as an undergraduate, I made arrangements with Dr. Wishart in Cambridge to pursue a career that would possibly have included doing a PhD with him but, sadly, the summer I talked to him he died in a drowning accident and as a result I ended up as a student of R. A. Fisher's.

Mathematics was a great preoccupation with many of the responses received as a result of my letter. Because it forms the basis for the physical sciences, perhaps that should be expected.

Bugs in Balloons

A high school speech request played a role in the professional development of Sir Richard Southwood, head of the Department of Zoology at the University of Oxford (UK). From the age of ten, Southwood had been interested in collecting insects and in high school was active in the school's science society. It was in conjunction with his duties as secretary of that society that he sent a letter to the director of the Rothamsted Experimental Station (Harpenden) to seek a guest lecturer for his high school. The volunteer speaker turned out to be C. G. Johnson, who was impressed with the tour provided by young Southwood. Johnson invited the student to participate as a voluntary worker in the Entomology Department at the Rothamsted Experimental Station. Southwood's first task was to help sort the insects caught in traps flown from balloons, and he soon noted that the relative abundance of species found in the upper air was different from those he had found in ground-level collections. Johnson was intrigued with Southwood's findings, and the two coauthored a short paper. In subsequent studies Southwood fostered the concept that the features of an organism's life history "had been forged on the anvil of its habitat."

An Attraction to Basic Research

Unplanned circumstances led Robert E. Berne, chairman, and Charles Slaughter, professor of physiology at the University of Virginia School of Medicine (Charlottesville), into his current field of investigation. He decided to become a cardiologist following medical school, clinical training, service in the Army during World War II, and a medical residency. He decided to bolster his background with a six-month fellowship in cardiovascular physiology with the "dean" of cardiovascular physiology, Dr. Carl J. Wiggers, at Western Reserve Medical School in Cleveland, Ohio. This exposure hooked Berne on basic research, and he accepted a faculty appointment in the physiology department there, thus starting an academic career in physiology.

Chance played a role in his research accomplishments in that adenosine, which was known to have properties compatible with a role for local regulation of blood flow, actually turned out to play that role. Adenosine's importance exceeded that which Berne and his colleagues had attributed to it.

A Ten-Second Window of Opportunity

Timing can be all-important. As a perfect ending to this chapter, an example of perfect timing is in order. In the following account, taken from the *Annual Review of Physiology (1985),* Horace Davenport tells of the chance occurrence that changed his life:

> In the spring of 1941 I was in my second postdoctoral year as a Sterling Fellow at Yale (New Haven, Connecticut). I was desperate for a job, having haughtily turned down the University of Chicago offer of an instructorship at two thousand dollars per year. At the Federation (of American Societies for Experimental Biology) meeting in the Stevens Hotel in Chicago, my current chief, C. N. H. Long, went through a door just as Cuthbert Bazett of Pennsylvania was about to go through it in the other direction. Bazett said 'Hugh, do you know anyone who wants a job?' Hugh Long replied 'Yes, that tall man over there.' Bazett had hired an Englishman who suddenly couldn't come to the States because of the war, so he hired me almost on the spot. Bazett was an Oxonian; I was an Oxonian, and therefore obviously qualified. So I became a physiologist. Ten seconds either way for Long or Bazett going through that door, and I would have remained a biochemist. Considering the postwar explosion in biochemistry and my inability to learn new methods, that was the luckiest thing that ever happened to me. Otherwise, I might now be an embittered associate professor in some intellectual backwater.

In the course of time he spent many years at the University of Michigan Medical School, chair of physiology, where he gained fame for his research on gastroenterology. His great rapport with his students was evident in that he built a wing on his house to be used for regularly scheduled gatherings. Davenport's favorite saying was "chance favors the prepared mind," and it was so impressed on his students that the signs used to denote the direction to picnic sights for the group bore the message: "Chance favors the prepared mind."[1]

REFERENCE

1. Davenport, Horace. 2003. *Medicine at Michigan*. Spring.

Chapter 13

Keeping the Monkey Wrench Out of the Gears or How Not to Screw Up Research

"I construe my function as a director of research as mainly to create the kind of environment which is conducive to the advancement of learning. That sounds pompous, but this is all a director can do. You cannot *direct* people to have ideas, and no one can have a big enough grasp of the whole of biological science to be able to say which lines of research are certainly going to be fruitful and which are certainly going to be a waste of time. So what one has to do is simply to create an environment and an atmosphere in which science flourishes."

—Peter Medawar

Medawar's attitude toward directing research is the epitome of wisdom according to the dictionary definition—"a wise attitude or course of action." Other philosophical observations by prominent research directors are worth noting. Marinus Los, research director of plant sciences at American Cyanamid Co. (Princeton, New Jersey), was honored for his development of the herbicides known as imidazolinones. Four of these products accounted for sales of more than six hundred million dollars in one year, the payoff for fourteen years of research and six to eight years of development The cost was thirty million to fifty million dollars for each product. In his address at the award ceremony in Williamsburg, Virginia, in October 1994 Los stressed the need for a creative scientist to optimize his chances for discovery by "looking for, expecting, and exploiting anomalies."[1]

Sir Gustav Nossal, a Nobel Prize winner, authored a quotation that should be required reading for every research director:

> In most discoveries as they first emerge, there is sufficient element of doubt and tentativeness to give scope to a stern critic; it is relatively easy for an intelligent man to pick holes in the incomplete but truly new discovery. The overly critical, highly intelligent, but unoriginal scientist can become a negative force in research.

One of the more accurate and incisive observations on research was that of Nobelist George Hitchings: "Individuals should be given maximum freedom in research because relevance can only be determined by hindsight."

Experienced researchers can relate to each of these observations. Probably many laboratories could serve as role models but the emphasis here will be on the one known best to the author, the Naval Research Laboratory (NRL) in Washington, DC. What made this laboratory special was its outstanding chain of command. To repeat Peter Medawar's sentiment, it created "an environment and an atmosphere in which science flourishes."

Chain of Command

In 1956 the Civil Service Commission made a survey of government laboratories in the Washington area and found that the lowest turnover rate was at NRL. Furthermore, the lowest turnover rate within NRL was in the chemistry division, which I was joining at that time. The satisfaction among NRL's scientists was definitely not attributable to inflated salaries, because at that time government pay scales lagged considerably behind those in industry. Nor was it modern equipment. The buildings were old, the air conditioning was practically non-existent, and the atmosphere was polluted by the neighbor next door—the Blue Plains sewage treatment plant. It is the largest sewage treatment plant in the world (often exceeding four hundred million gallons per day). On the positive side there was adequate funding for sustained basic research in many areas.

A major factor in the morale at NRL was its chain of command. At the time I came on board my immediate superior was John Leonard, whose PhD degree was in biochemistry. His grasp of all of the physical sciences was greater than most of the scientists I have known. In addition, his keen interest in the use of English made him invaluable as a report writer or editor. John organized what he called the "Sesquipedalian Literary and Chowder Society," which met regularly to discuss whatever multisyllable words its members had encountered.

Our branch head was Allen Alexander, one of the NRL contingent of organic chemists who received their PhDs at the University of North Carolina. "Alex" began his career as a paint chemist with the Sherwin Williams Company, and his duties at NRL concerned many aspects of coatings for which the Navy had a need. In fact, this particular branch came into being because the flying boat *Mars* had been unable to take off when the heavy coating of barnacles on its hull reduced its speed so greatly. Alex also served as associate superintendent of the chemistry division and was largely responsible for handling personnel-related matters in the chemistry division. He had an engaging personality that fostered many contacts within the Navy structure and thus helped in locating significant funding for various research projects. In 1960 he was elected president of the local section of the American Chemical Society.

Dr. William Zisman was head of the chemistry division and his technical expertise is described in detail later in this chapter. Surface chemistry was his abiding interest and everyone in the division was trained to be aware of the various interactions that can be involved in diverse natural systems. As an example, he had a great influence on Bob Baier who served as a postdoc in Zisman's laboratory from 1966 to 1968. Prior to his advanced training, Baier had developed an interest in medical disciplines, and Zisman, recognizing his enormous potential, urged him to bring to the medical world an awareness of the inherent advantages of surface chemistry techniques. Accordingly, Baier became associated with IUCB (Industry/University Research Center for Biosurfaces) at the University of Buffalo (New York). He became biomaterials graduate program director and continued his collaboration with Zisman long after he left NRL. Following Zisman's death in 1986, Baier published an article in the *Journal of Dental Research* entitled "William A. Zisman: A Scientific Pioneer in Dental Bonding" in which he described the scope of Zisman's work. Among Zisman's awards was the Hillebrand Prize from the Chemical Society of Washington, the Kendall prize awarded for research in colloids, and a festschrift volume of the Advances in Chemistry series (Volume 43, "Contact Angle, Wettability and Adhesion" dedicated to him in 1964). Baier recalled that "As his postdoctoral apprentice in 1966, '67, and '68, who as late as 1976 (when Dr. Zisman was still calling about overdue manuscripts) would automatically rise from his chair and stand at attention while talking to the man over the phone, five hundred miles away." Zisman published over one hundred fifty scientific journal articles and held thirty-six patents.

One step further in the chain of command was Dr. Peter King, the associate director of research for materials, of whom much will be said later in this chapter. King's most noteworthy accomplishment was the role he played in pinpointing the exact time that Russia set off its first atomic blast.

The director of research in early 1956 was Oscar Martzke, who left shortly thereafter to become vice-president for fundamental research at the U.S. Steel Corporation. He was succeeded by Dr. Robert Page, who had spent many years of research at NRL in the field of radar. Page held more than fifty patents in radar.

This litany of accomplishments demonstrates the technical expertise found in every level at NRL. Employees throughout the laboratory had every reason to respect the judgments made by such a group of outstanding people.

Cooperative Spirit

Approximately three thousand full-time employees represented a broad spectrum of scientific disciplines and shared an unusual degree of cooperation. No matter what hurdle loomed, there seemed to be someone within NRL who was familiar with that particular field and was willing to help with either advice or equipment. Also there was an uncommon bond between machinists in the various shops and the technical staff. Included in this mix of people were interesting types who lacked the formal training required of scientists, but had innate abilities that were vital. In one session of reminiscences about the chemistry division, Kingsley Williams paid tribute to several people "who could get things done in circumstances where the eggheads of the group were absolutely helpless."

Peter King

Luther Lockhart

One of the most noteworthy cooperative studies among different divisions featured Peter King and Herbert Friedman, who was elected subsequently to the National Academy of Scientists. It began in April 1948, when Friedman and Irving Blifford, both physicists, were using Geiger tubes to detect changes in atmospheric radioactivity following the U.S. series of nuclear explosions at Eniwetok. They found an increase in radioactivity after a rainfall, but they couldn't identify it as fission product radioactivity from an atomic bomb. Friedman asked King if there might be a possible chemical approach to identify the radioactive products scavenged from the air by rain. King reasoned that it might be possible to concentrate the radioactive particles by the use of aluminum hydroxide, such as that used in water purification assemblies. King brought Luther Lockhart into the project and asked him to consult with the staff at the Dalecarlia Water Treatment Plant in Washington. They instructed him in the process of aluminum hydroxide flocculation to trap minute particles, which Lockhart then used successfully in trapping dissolved ^{60}Co from a gallon of water

The next step for Lockhart was to visit the Virgin Islands site where rainwater had been collected in large cisterns during the Eniwetok atomic explosions. With the cooperation of the associate superintendent of the chemistry division, Mr. Gulbrandsen, two thousand five hundred gallons of water were treated to give a concentrated slurry totaling just five gallons. How to deal with this melange of chemical species was not known because such technology was reserved to only those who had been directly involved in the Manhattan project. While they were able to get an inkling of how much of each isotope was formed in the fission of ^{235}U by reading the Smyth report that was released after WWII, they still had no information on the fission products from either uranium or plutonium in a nuclear explosion. In a personal recollection written recently Lockhart said:

> Gulbrandsen and I worked out a scheme for the chemical separation of the critical fission products, while Friedman and Blifford worked out procedures (and developed equipment) for measuring the half-lives and the energies of the emitted beta particles. Anyway, we were able to separate and identify positively several of the radioisotopes and, by extrapolating backward to a time when their radioactivity ratios corresponded to those expected of a nuclear detonation, we were able to give an approximate date of the tests. All of this was done within six months of the time of Friedman's initial observation!

There were subsequent tests made of water collected at St. Thomas, Truk Atoll, and Hemya, Alaska, at the tip of the Aleutian Islands, to determine whether any radioactivity from the U.S. tests had reached those sites. Of particular interest was the Aleutian sample because it confirmed that there had been no prior Soviet test.

An interesting sidelight to this development was that the report Lockhart and Blifford had written was declared to be top secret, which meant that they, the authors, did not have the clearance to read it! Then the detective work began in earnest. Dick Baus, a recent addition to the chemistry division, and Lockhart collected rainfall routinely from a section of roof on a building at NRL, treated it with aluminum hydroxide floc, and looked for radioactive fission products. What they found were quantities of natural radioactive materials (from radon in the atmosphere) with traces of fission product activity. It was possible that it had drifted from Oak Ridge (Tennessee). They tried to get a "clean roof" by scrubbing the tarpaper with a detergent and having the fire department hose it down. After treating the water as they had done before, they found "really enormous quantities" of the natural radioactivity together with more fission products. Similar results were obtained with collections made at NRL's Chesapeake Bay Annex. Then, on the basis of balloon measurements of radioactivity that had been made by Friedman's group during the Eniwetok test, it was concluded that these fission products were deposited during the passage of a radioactive cloud over NRL in early 1948. In any case it was evident that a new roof was needed if they needed a contaminant-free surface.

By early summer 1949 a new 1000-square-feet aluminum roof, new gutters, and a collection tank capable of holding several hundred gallons of rainwater were in place. And they were collecting and analyzing every rain! They had also equipped the Naval Station at Kodiak, Alaska, with a similar facility. Naval corpsmen there would collect the rain, treat it, and return the concentrated floc to NRL for analysis. Both sites were also equipped with air filtration devices, developed by Friedman's group, which filtered air over a twenty-four-hour period, for radioactive analysis. When the Soviets set off an atomic bomb, the NRL team was ready to detect it. Rain fell during the passage of the radioactive clouds, and delivery to NRL from Alaska was expedited by a special Navy flight. Even the chemical procedures developed by Gulbrandsen and Lockhart worked, despite the fact that they encountered a number of short-lived fission products they had never faced! Following is a quote from the introduction to the TOP SECRET RESTRICTED DATA NRL Report, dated 22 September 1949:

> Positive radioactive evidence of a recent explosion of an A-bomb has been accumulated by NRL fission detection stations at Kodiak, Alaska, and Washington, DC, during the period from 9 September to 20 September. The date of fission, deduced from radioactivity ratios of fission isotopes, is probably not earlier than 24 August. Extremely hot samples extracted from the fallout of fission products at Kodiak have yielded tens of thousands of counts per minute

> of the major fission product isotopes. This report is a brief account of the methods of detection and the fission activity measurements completed to date. More detailed reports are now being written ...

President Truman announced to the world on September 23, 1949, that the Soviets had exploded a nuclear bomb on August 29. All of the NRL participants were given special awards (in-grade promotions), and later Pete King and Herb Friedman received major Navy and Defense Department awards based on this work, which was totally memorable.

Notable collaboration between divisions often featured work done by Jim Griffith, a brilliant paint chemist, who spent thirty-six years at NRL. When the United States decided to build a "stealth" airplane, one of the needs was for a coating that would absorb radar frequencies. This coating also had to be flexible in the range of temperatures to be encountered by the aircraft and must be able to adhere to the aircraft surface. Griffith was able to achieve those goals.

Project VANGUARD was the forerunner of NASA, and among its needs was a detector for micrometeorites because there existed the possibility that a fast-moving particle, though having a small mass, could destroy a spacecraft. The satellite NRL was developing was made of a magnesium alloy and the welding of the chambers was not perfect. A sealant was needed that could go into such a chamber and hold its pressure for some period of time sufficient to determine whether it had encountered a micrometeorite. Because of the severe weight limitations envisioned for the satellite, they asked Griffith to limit the addition to only four grams of sealant. He stopped all of his normal work and developed a liquid polymer that could be injected into the chamber, after which the satellite would be rolled around, and the excess could be pulled out. This liquid polymer would find its way into the leak site, seal it, and then solidify. His success in this venture led to more requests from the VANGUARD personnel. One of them had devised a micro-battery and was concerned that it would spill its contents once it got into the hard vacuum of space. The microelectronics in those days were flimsy, so Griffith devised a polyurethane foam that would encapsulate the foam. Griffith was quoted as saying:

> In those days there was a lot of not only intra-, but also inter-division free help. In other words, none of us ever thought of asking for a job order to do these side jobs. It was just that there was a super interesting project over there that was having a problem, so we picked up and helped them.

When the Navy began to develop a Fleet Ballistic Missile (the Polaris program), the matter of weight conservation was critical. For a rocket to be fired from a submerged submarine, its casing had to be as light as possible. Metal casings were discarded in favor of cases made with filament windings, which were made from reinforced epoxy resins that were wound on a mandrel and then cured. One problem with the commercial products used was that they had a tendency toward premature damage during proof-testing (i.e., craze cracking in the resin). Griffith was experienced in various resins and devised one that would resist craze cracking. He contributed regularly to the Polaris motor case conferences, which were at Aerojet General and Lockheed in California. When he retired from NRL, he had a long interview with the resident historian, consisting of seventy-three pages. The examples given here are only a few of the many discussed in that long interview.

Things could be done in a hurry at NRL. Someone had a problem with lubricating a certain system (some of the particulars have been lost) because the silicone that was tried did not work. When told what the particular problem was, Bob Fox had a hunch that dodecyl sulfonic acid would do the trick, and he happened to have a small amount in his laboratory. Sure enough, it worked, but how can some more be gotten? Bob knew that a friend of his at Columbia Organic Chemicals in South Carolina had on hand a large amount of dodecyl bromide which could be easily converted to the sulfonic acid. Forthwith a Navy captain flew to South Carolina, picked up a kilogram of the dodecyl bromide, and brought it back to Fox's laboratory. Because so little was needed for the job at hand, it is likely that there is still dodecyl sulfonic acid available for use.

Gyroscopes and other rapidly spinning machinery present problems because the lubricants used with them can spin off. Once again, a silicone lubricant did not provide the protection that was needed. To the rescue came Jim Romans and Vince Fitzsimmons, who were familiar with a "barrier film" that was devised to have a "critical surface tension" (another surface chemistry application). They lubricated some very small bearings in a guidance device and placed the barrier film around the assembly to prevent the lubricant from spreading. The device was then sent to a testing laboratory to determine whether it met the specifications for a three-hundred-hour service life. Weeks went by and they had not heard from the testing laboratory. Finally, the testing laboratory called and asked, "How long do you want us to run this test? This thing's been going for three thousand hours and still hasn't stopped!"

Many more examples could be given but these illustrate the types of relationships that were required in a laboratory in which so many diverse scientific disciplines were studied.

William Zisman

Recollections of the Chemistry Division

A balance between basic and applied research is necessary for an organization to reach its potential, and William Zisman, head of the chemistry division, recognized that balance. He had been trained as a physicist under Percy Bridgman, a

Nobel Prize winner. His formative years coincided with the Depression, when funding for science was extremely scarce. Consequently he worked for a significant amount of time at no pay with the Carnegie Institution's Geophysical Laboratory to get the experience he needed. One of the early papers he published, with Harry Fox, bore the esoteric title "Some Advances in Techniques for the Study of Adsorbed Monolayers at the Liquid-Air Interface," which was the first in a series on surface chemistry phenomena that provided solutions to diverse Navy needs.[2] For example, when the USS Constellation suffered a major fire while at its berth in Brooklyn, New York, one of the consequences was a huge loss of electronics equipment because of its immersion in oil and salt water. In 1960 dollars the equipment was valued at more than six million dollars and the time required to order replacements would be six months. But, thanks to a basic understanding of oil/water displacement systems developed by Zisman's group, most of the electronics gear was salvaged in a few weeks.

It might be a surprise for many to know that Zisman played an important role in the development of the Teflon frying pan. Teflon was discovered in 1938 at DuPont (by Roy Plunkett) but the Teflon pan was not marketed until 1962. Initially DuPont's hope for Teflon lay in its inertness, as illustrated by the remark of a DuPont researcher at a national meeting of the American Chemical Society: "If you develop a universal solvent, we've got something you can store it in." Zisman was a friend of the director of research at DuPont and, during a visit this gentleman made to NRL, Zisman pointed out that the more important attribute of Teflon was its unique non-wettability. That led to the application of Teflon to frying pans as a nonstick surface.

Zisman cooperated with DuPont in 1948 in a study of Teflon's remarkably low coefficient of friction. Ten years later, in a paper by Vincent Fitzsimmons and Zisman, they showed the effectiveness of Teflon as a dry-film lubricant for gun cartridges.[3] A paper in 1968 by Bowers and Zisman[4] contained the following:

> (Other) examples are the common uses of dry-film lubricants in space vehicles, rockets, inaccessible portions of submarines, and in food processing, packaging, and textile industries. Especially advantageous is the permanence of dry films during storage and their ability to function over a much wider temperature range than oils or greases.

All of these advances are instances of practical applications arising from basic research.

Not long after the first nuclear submarine joined the fleet, representatives of academic and industrial laboratories suggested a back-to-nature approach for maintaining a habitable atmosphere within the ship. Theoretically, a mass culture of algae could absorb the CO_2 that was exhaled by the men and replace it with O_2. Already available and working well was an electrolysis system for producing oxygen and an ethanolamine scrubber for removing carbon dioxide. The suggestion for adopting the algal system was based on the premise that it would achieve both of these goals, and in addition might also scrub out some of the kitchen, bathroom, and cigarette odors.

The Office of Naval Research (ONR) had sponsored research on algal systems in several institutions. However, it felt that NRL should provide an impartial view of what could be expected from mass cultures of algae. John Leonard had done a feasibility study that indicated the algal approach was worth investigating. NRL searched within its ranks for a chemical engineer to undertake the task. No such person was available and in this vacuum of scientific expertise, I was tapped for the job. This was essentially by default because my formal training had been in organic chemistry, and I had no training in biology. Two other chemists, Constance Patouillet and Bob Shuler, were also assigned to the project and their lack of suitable background was the same as mine. In the early stages of this project, I ran into Dr. Zisman in a hallway and his words were both encouraging and challenging: "Jerry, I want this to be the best study of its type in the country." What was important was that he set no time limits on the study, nor did he ever attempt to micromanage it. Subsequent developments showed the wisdom of such an approach.

I was well aware that at least one industrial laboratory had already been working on the problem for more than a year, with more than ten people (including PhD phycologists) assigned to it. It was not reasonable to expect that an inexperienced group of three chemists could match the efforts of the larger industrial group having an educational background more suited for the task at hand.

Gradually, I learned a great lesson about the necessity of freedom in pursuing research goals. The industrial laboratory with funding by ONR had made its entry into the field by hastily constructing a photosynthetic gas exchanger with inflexible characteristics, before it had an opportunity to determine what the critical factors might be. My assistants and I were ignorant, but we had the luxury of being allowed to explore the importance of each variable (e.g., light intensity, rate of addition of fresh culture medium, CO_2 concentration, etc.) before committing ourselves to a design. We were using the same organism as the industrial laboratory. But what became clear as a dominant factor in the rate of oxygen production was the distance between the light sources. This had not been explored by the industrial laboratory because of the commitment to its first design, in which the lights were comparatively far apart. When we placed the lights close to each other, it was clear that it was the key to high productivity. The rapid, pragmatic, design used by the industrial laboratory in constructing its gas exchanger was a poor choice, and the rate of oxygen production they achieved was far less than ours.

Lest the reader wonder about the outcome of the project, the results can be summarized briefly: oxygen production rates by the algal culture were extremely reproducible, but the electrical power requirement for the incandescent lights was excessive.[5] For a submarine with one hundred men aboard, the algal culture would require more electrical power in one day than a three-bedroom house would use in fifteen months.

I cannot let an opportunity pass to state once more that new approaches are not always welcome in the scientific world. Measuring O_2 production and CO_2 absorption rates consisted in bubbling input air through the culture at a measured rate, after which the air went through a drying column and both an O_2 analyzer (paramagnetic) and a CO_2 analyzer (infrared). The O_2 production rate, for example, would be computed as the product of the airflow rate times the fractional increase in O_2 concentration. It would be an instantaneous measurement. This procedure is known as the gas exchange approach, which is very unique in that *each measurement is a rate measurement.* Consequently, one can determine *instantaneously* the effect of a change in culture conditions on the growth rate of the culture when, say, a toxicant is introduced. There are extremely few technical reports that are based on this phenomenon despite its simplicity. By the time I was nearing retirement from NRL, I had become frustrated that the scientific world was not impressed with this (what had become to me) obvious effect. So I called Dr. Jack Myers at the University of Texas, whom I considered the most knowledgeable man in the field, to ask him why this might be. I had met Dr. Myers several times, and he was somewhat familiar with my work. When I asked him the question, he paused for a significant amount of time before answering. Then he said, "Well, Jerry, you're a chemist and you think like a chemist. These people would rather look through a microscope." He's probably right.

One more story must be told about Zisman that demonstrates his dedication to public service. At a meeting when budgets were being discussed, a project leader made a casual remark that the cost of a particular study was "estimated at $25K." Zisman stopped him abruptly: "I presume you mean twenty-five thousand dollars, and to me that represents a lot of money. The term 'K' will not be used in these discussions."

One of the branch chiefs in the chemistry division deserves special mention for the range of contributions he made to the Navy and to the public. Homer Carhart served fifty-two years at NRL from 1942 to 1994 and received almost every award the Navy could invent. As head of the fuels branch, he directed research of a basic nature that impacted on Navy practices in the handling of aircraft and ship fuels. Studies in his branch included cool flames and precise measurements (by Will Affens) of flammability limits of various fuels. Subsequently, Carhart was given additional responsibility for developing fire control methods in surface ships and submarines. One of the innovative efforts undertaken by his group was a study of the feasibility of snuffing out fires in nuclear submarines by the rapid release of nitrogen from large pressure tanks. Crew members can sustain life while breathing air containing only twelve percent oxygen; that is below the concentration required to sustain combustion. At this writing, the methodology has not been adopted for fleet use but the chances are good that it will ultimately be approved.

Jerry Hannan and Homer Carhart

In 1986 the Navy Technology Center for Safety and Survivability was created at NRL, with Carhart as head. Under his leadership this center provided science, the Navy, and the nation with an astonishing flow of innovative discoveries. Virtually all personnel and every ship and aircraft in the Navy have been made safer and have a greater chance of survival in a fire. Some specifics:

Research on fuel properties and ignition behavior led to the adoption of his recommendations for new flash point standards for Navy and civil aviation fuels. It was found that flash points could be reduced from 140° to 125° C by the addition of only a small amount of the lower flash point material. In the 1970s, Homer was asked by the Federal Aviation Administration to chair a blue ribbon panel on airplane safety with regard to the fuels used. After studying the results of many airplane crashes, they were somewhat surprised to find that there was a definite saving of lives depending upon the flash point of the fuel in the plane; the higher the flash point, the fewer deaths by fire. When this became known, the airlines adopted mandatory rules regarding the flash points of the fuels they were using.

Hazards related to the generation of electrical charges in the transfer of fuels and consequent explosions were studied and control measures were adopted.

A handheld infrared imaging device was developed, allowing firefighters to see through the blinding smoke of shipboard fires.

Firefighting techniques in the Navy and in worldwide civilian applications have been improved through the development of an aqueous film-forming foam (AFFF) and PKP dry fire-fighting chemical (Author's note: PKP stands for Purple K Powder which is essentially potassium bicarbonate). For example, fire trucks in rural areas are forced to carry their own water, but with the use of AFFF they can extinguish a fire with half the amount of water normally required.

Communications throughout a ship that is on fire have been made possible by the development of an acoustic system embedded in the hull.

Carhart's leadership skills started with his personality—definitely a "people person." He had friends throughout the Navy and that served to provide him with funding for whatever project he felt was needed. Those in his branch felt very close to him, and everyone knew him as "Homer." It was my pleasure to serve in his branch for several years before my retirement, and I came to know him well. Homer is an enthusiastic musician and throughout his years at the laboratory composed many works for the piano. Upon his retirement, his family presented him with a special computer that actually prints the music being played on a keyboard, making it possible to preserve his compositions for posterity. When I became aware of this, I asked Homer if he would assist me in putting on paper a composition that had been running through my head for many years. Not only did he do that—he provided a beautiful obbligato for it that would never have occurred to me.

Attributes to Be Found at NRL

In 1973 as a part of its fiftieth birthday observance, NRL published a book of reminiscences of many of its distinguished personnel.[6] Glimpses of these writings illustrate the breadth of studies conducted there, and the general atmosphere that prevailed. To begin, here are some of the reminiscences by Dr. Peter King whose work has already been mentioned. He came to NRL before the start of WWII, and here he describes what it was like during the early days:

> "If you hear any noise around here, shoot first and then go see what it is." One evening very early in World War II, when several of us at the Lab. were staying a bit late because we wanted to, I heard those directions being given by a Marine sergeant to a member of that night's guard force. There was a break in the fence next to the building, due to some construction, and they were examining the break just below me, unaware of my presence. It obviously was time to leave before it got too dark, seeing that we would undoubtedly make some noise close to that break when we left, and we wanted safe passage! Despite the humor of the situation there was something else more significant. Somebody had told the "Sarge" that the work going on was important to the Navy and the nation, and he believed it, thus joining a small club. He was giving physical protection and other kinds were also needed—financial continuity, and freedom of investigation, and these too were being protected by the then—also small group of civilians and officers who were in a position to do it and believed in the Lab.
>
> "If I hadn't had the FTS (False Target Shell), I would have lost the boat and crew." How does one react to a statement like this in a war patrol report by a submarine skipper? I have seen the reaction and what it is is great satisfaction and the encouragement to try something else new. The feelings of the individual really cannot be described, but at least he can now go on even if what he is working on at that moment is equally foolish in the minds of the uninitiated.
>
> "Without the re-breather we would have lost the ship." This statement was made in reference to a re-breather used in a very bad fire aboard one of our carriers. When reading this, one can forget all the extra effort and time that went into setting up a small pilot plant to produce in quantity a superoxide

that was just a curiosity a short time before. (The superoxide was the key ingredient in the Oxygen Breathing Apparatus, by providing oxygen and also removing CO_2. The pilot plant was in the Chemistry Building at NRL).

How do you convey the feeling of the place and the people in it, to the unbelievers who have not been a part of it? How do you make them understand that this history, this desire to do something, is not just of the past but is now going on with different people and different problems and that "protection" is still needed?

What was it that made people drop into the Lab. on Saturdays during mid-1940s, and up through the first eleven months of 1941? It was not money, for they did not get any for doing it. It was not for glory, because there was none of that either. We were not at war, so it was not that either. It was done because people saw things they knew needed to be done, and they wanted to do them. There was an atmosphere of urgency surrounding most of the NRL people, which made them do what they thought was needed. As time passed, more and more scientific types appeared on Saturdays until finally one would believe from the number of cars that it was just another normal day. This was the spirit of the place then and still is.

VANGUARD—Venturing into Space

After World War II, research at NRL included airborne instrumentation for the study of the earth's atmosphere, using rockets captured from the Germans and also newer models developed as part of the program. This resulted in the formation of the Atmosphere and Astrophysics Division. With the advent of the International Geophysical Year (IGY), this division offered to take part in the IGY Earth Satellite Project. NRL won funds to start the program in October 1955. Because of the diverse needs of this program, a group outside the division structure, called Project Vanguard, was organized and headed by John P. Hagen. A portion of his summary of Project Vanguard is shown below:

The Vanguard team's first pressing problems were to properly organize for the difficult task and to immediately get work started on the design and construction of the vehicle. Experience in the Department of Defense at that time showed that in missile programs it took more than five years from the start of a program to arrive at the date of the first successful launching. Vanguard had three years to launch the satellite in the time frame of the IGY. The team achieved the objective in two years, six months, and eight days. To make this possible, groups were organized to develop a suitable launching vehicle; to develop a scientific program which would design the satellite; to miniaturize the instruments to be placed in space and arrange to telemeter the observa-

> tional results back to earth; to develop a worldwide system of tracking which would be capable of monitoring the progress of a multistage vehicle during launch and then to precisely track the satellite while in orbit; to find and instrument a suitable location for launching the vehicle; to provide liaison with and to coordinate the work in other parts of DOD in support of the Vanguard effort; and to help in administering the project through the development of a fiscal plan and managerial plans to ensure adherence to schedules.

This major effort resulted in the successful launch on March 17, 1958. In time it was recognized that there should be a totally new structure for satellite research, and Project Vanguard was moved to Beltsville, Maryland, where it was incorporated into the National Aeronautics and Space Administration (NASA).

From Sea to Shining Sea

One of the strengths of NRL in its early years was in the study of optics. E. O. Hulburt headed the optics division and then served as the director of NRL for many years. A measure of the breadth of these studies is to be found in the remarks of John Sanderson, who retired from NRL as head of the optics division in 1965 and served later as president of the Optical Society of America:

> When I joined the old Physical Optics Division in 1935 the Naval Research Laboratory was only twelve years old. But optics had already become something of a sea-going division, along with several others whose work related to the operations of the Navy on and under the sea and in the atmosphere of the earth.
>
> As a part of my welcome as a member of the NRL, E. O. Hulburt told me about the policy he had set for himself when he became the first superintendent of the optics division on June 4, 1924. (It was then called the heat and light division: I will call it the optics division with the certain knowledge that this title will be understood). He said that he thought a naval optics program should include subjects of general interest, not under thorough cultivation elsewhere, which would lead to results of interest and potential usefulness to the Navy.
>
> These introductory conversations also left no doubt in my mind that I was to consider myself a member of the NRL and the Navy, and not merely an employee or hired hand. That was the doctrine throughout NRL, and certainly it has had a profound influence on the long-term strength and stability of the laboratory.

An appreciation of Hulburt's renown was evident in a literature search which I made in the mid-1970s. In studying the deleterious effects of ultraviolet light on marine algae, it became necessary to know how far the UV penetrated into the sea. A range of absorption coefficients for UV could be found in the literature, but I came across (and cannot now locate) an article in which the writer had analyzed critically the manner in which they had been obtained. According to his view, the most accurate absorption coefficient was obtained by Hulburt and Dawson in 1936. This predated the phototube, and Hulburt and Dawson had used the darkening of a photographic plate as a measure of light penetration!

Deep Sea Research

Finding sunken submarines and other material (such as an H-bomb) is a task requiring unusual equipment and capabilities. One of those who contributed greatly was "Buck" Buchanan who joined NRL as an electronics engineer in the sound division and eventually became head of the ocean engineering branch of the ocean technology division. He was one of those who converted an ice-breaking freighter into a research ship (*Mizar*) with a unique center well through which various sensors and "fish" could be deployed.

When the submarine *Thresher* sank with a total loss of life, there was great concern about the cause. An examination of the wreckage was a high priority, and a search effort was carried out to combine the features of *Mizar* with the mini-submarine *Trieste,* which was the research vessel of the Navy Electronic Laboratory in San Diego. Buchanan's description of the search for *Thresher* included these words:

> After a week or so, we were confident of our ability to "work" the ship. We headed back into Boston to make our final preparations.
>
> Meanwhile the *Trieste* had been brought to Boston and was also being prepared. Since her preparation involved the safety of men, the work was more exacting and she was not ready for sea. Admiral Coates was the task force commander, and Captain Frank Andrews was task group commander. Their fleet consisted of the *Mizar,* the *USS Hoist,* and the submersible *Trieste. Mizar* sailed on June 25, 1964, to start the *Thresher* search. In the incredibly short time of two days we located and photographed the hulk. By July 22 we had photographed almost the entire hulk, which was broken into five major pieces.
>
> Proudly, yet sadly, we sailed into Boston on July 23 with a broom displayed at the signal halyard (author's note: the traditional Navy symbol that the sea had been swept clean).

By this time the *Trieste* was nearly ready, and we all turned to in preparation for the task of guiding *Trieste* to the wreckage to make a close-up inspection.

In early August the entire task force arrived in the area, and preparations were made for *Trieste* to dive. Her first dive on August 14 was unsuccessful, due to a fault in her gyrocompass. Her second dive also was thwarted by the same problem. During these dives we on *Mizar* were gaining experience in tracking her and were very sure that once her gyrocompass was operational, we could direct her with ease.

The third dive proved to be one of the most amazing operations we could have imagined. Captain Andrew elected to make the dive as observer. The officer in charge and pilot were Lt. Commander J. B. Mooney and Lt. John Howard, respectively. As the dive progressed we all breathed a sigh of relief as it became evident that the gyrocompass was working. For a short time the *Trieste* rested on the ocean floor at eight thousand four hundred feet, while we meticulously determined her position by use of the UTE (Underwater Tracking Equipment). Lt. Denny Curtis was the dive director aboard *Mizar,* and he was frequently in contact with those aboard *Trieste.* Unfortunately, the telephone transmissions interfered with the UTE operations, so that we had to restrain these communications.

We informed Captain Andrew of the course to steer and the distance to the *Thresher* hulk and, wonder of wonders, the *Trieste* position moved slowly in the correct direction.

This had never been attempted before, and we really had no way to determine the accuracy of the system. Finally, the little dots on our plotting board reached the point where we had previously judged the *Thresher* hulk to lie. Lt. Curtis informed Capt. Andrew that they were now so close that they were within our circle of confusion. Now Capt. Andrew is an excellent mathematician and a great believer in statistics. He therefore asked the pilot to set *Trieste* down on the ocean floor. He then informed Lt. Curtis that they saw nothing, and requested that we compile a lot more data and average it so we could vector them closer to the hulk. We did so, but to no avail. The statistics always converged on the spot we had selected (and which by now we had begun to question).

Trieste is a pretty tight fit for three men. The third man, who is looking out of the view port, is sometimes almost forgotten down on the bottom of the dark capsule as he peers out into the murky water. As they sat still waiting, the observer said, "Say, how high are we above the bottom?"

"Man, we're resting on the bottom, we're just waiting for those clowns up there to tell us which way to go" was the pilot's response.

Observer: "Well, then, why can't I see the bottom?"

Pilot: "You can't see the bottom?"

Observer: "Heck, no, I can't see anything."

Pilot: "Wait a minute. Let me turn this thing around."

Observer: "Holy cow, there's a piece of metal —and it's made of metal, it's the hulk."

The *Trieste* had been setting directly on the hulk.

Buck Buchanan

While Buchanan's recollection in 1973 was in a somewhat jocular vein, that was not the case years before when I heard him give a lecture on the search for the

Thresher. The realization that they were at the site where so many men had lost their lives was so disheartening that his emotions almost prevented him from concluding the talk.

Mizar

One other aspect of the *Thresher* search should be mentioned here. The use of special strobe-type lighting for deep ocean photography, which was necessary for the task at hand, provided a unique look at the life to be found at eight thousand four hundred feet. The search group later developed the LIBEC system (for Light Behind Camera) that permitted taking very large area photos of the bottom. This bridged a gap in the scale of features on the sea floor between those too small to discern on sonar, and those too large to be contained in single conventional bottom photos. Wally Brundage, who had done his graduate work in oceanography at the University of Washington, was involved in the photographic work and had a full understanding of the unique opportunity, inherent in this extended search, to study life forms that may have never been seen before.

During most of 1963, conventional stereo camera systems mounted on towed racks were used. Late in that year, an improved camera system was used, having a strobe light mounted on the nose of a streamlined body, with two cameras mounted on the tail, tilted 17° from the vertical to each side to obtain wider coverage. Ultimately there were three Edgerton, Germeshausen and Grier (EG & G) cameras occupying the space between the tail fins. A complete description of the equipment is to be found in *Deep Sea Photography*. Serendipity played a role in providing a measure of the sizes of the organisms photographed. In the June 1963 studies, a number of light-colored sea urchins were photographed on or near familiar objects, such as a quart milk container and a Winston cigarette carton. From their size relative to the known objects (excluding spines), these urchins were determined to be four to five centimeters across. After that, the ever-present urchins served as a convenient scale, particularly in identifying the 1,769 kg mushroom anchor carried by *Thresher.*

Structures That Matter

To limit the mention of specific names to only a few eminent researchers at NRL is necessary lest this book be doubled in length. Obviously, however, there must be a reference to the work of Jerome Karle who, with Herb Hauptman, received the Nobel Prize. Their work covered more than twenty years of laborious mathematical calculations that were necessary for the interpretation of a crystal's structure, based on its reflection of X-rays. Even though NRL traditionally was oriented to basic research, there had to be remarkable patience among the various directors to continue Karle's work for such an extended period. Surely there were times when someone must have felt that this whole effort was a wild goose chase. Besides, what need was there in learning about crystal structure in such great detail?

Virtue prevailed and the Nobel Prize was awarded to Karle and Hauptman in 1985. Predictions about the greatest applications to be expected from this research would be fruitless, but one example will demonstrate the power of the methodology they developed. At the National Institutes of Health there had been great interest in the poison applied by Colombian Indians to their hunting arrows. It occurs naturally in the frog *Phyllobates aurotaenis.* The poison, given the name batrachotoxin, is the most potent nonprotein venom known. With great difficulty the scientists at NIH had finally purified a tiny single crystal of this toxin, just 0.03 mm in diameter. To determine the composition by normal chemical means might be impossible, but the crystal was large enough for Karle's group to work it out. What price tag do you put on such work?

An interesting commentary on this remarkable accomplishment is that it was not publicized lest the impression be given that NRL, as a laboratory of the Department of Defense, was involved in chemical warfare. A more accurate representation would be that NRL, devoted to science generally, would be glad to tackle an interesting problem without regard to its origins.

The remarks below are taken from Karle's recollections in the 1973 volume:

> Questions of policy concerning the support of science which profoundly affects the course and extent of scientific research arise anew from time to time. Currently the trend is to favor reduced support, with emphasis on short-term goals. The implication is that our society cannot afford to do otherwise. If one pauses for a moment to consider the consequences of such a policy on scientific and technological progress, both in terms of the creativeness and the productivity of the established professionals and the encouragement of young, talented, people to make the sacrifices of time and expense which are necessary for the development of their talents, one must conclude that a policy which offers reduced support and favors short-range goals, disproportionately, is most certainly the one thing that our society cannot afford—not even remotely.

A Personal Note

In this chapter attention has centered on the technical accomplishments within NRL, and also the atmosphere in which research is conducted. I close with a personal experience that occurred after having been at NRL only several years, when I was stricken with hepatitis. It was not a life-threatening situation but was totally debilitating, the result being that I had to take a month off from work. I was impatient to return and did so, probably before I should have. Within a week it became apparent that more time was needed for recovery, so I stayed home again for a few more weeks. During that time the associate director of research, Dr. Wayne Hall, sent a beautiful, long, handwritten note, urging me to stay home until I was certain that my health had been restored.

If this show of kindness has colored my regard for NRL, so be it.

REFERENCES

1. Industrial Research Institute. 1994. Fall meeting.

2. Fox, H. W. and Zisman, W. A. 1948. Some advances in techniques for the study of adsorbed monolayers at the liquid-air interface. *The Review of Scientific Instruments.* 19: 274–276.

3. Fitzsimmons, V. and Zisman, W. A. 1958. Thin films of polytetrafluoroethylene resin as lubricants and preservative coatings for metals. *Ind. and Eng. Chem.* 50: 781–784.

4. Bowers, R. C. and Zisman, W. A. 1968. Pressure effects on the friction coefficient of thin-film solid lubricants. *J. Appl. Phys.* 39: 5385–5395.

5. Hannan, P. J. and Patouillet, C. 1963. Gas exchange with mass cultures of algae. II. Reliability of a photosynthetic gas exchanger. *Appl. Microbiol.* 11: 449–452.

6. *Report of NRL Progress: Fifty years of science for the Navy and the Nation.* 1975

CHAPTER 14

A PHILOSOPHY FOR SCIENCE

"If you stopped the average man on the street, you would find that he had less than average intelligence."

—Dale Bultman

Most of the accounts in this chapter were obtained through the personal letters I sent to recently elected members of the Royal Society of London and the National Academy of Sciences in the United States. Also included are items uncovered in my diverse readings through the years. There are many scientific disciplines, but one common thread to be found among eminent scientists is that of curiosity.

Robert J. Lefkowitz, the James B. Duke Professor of Medicine at Duke University Medical Center, addressed the American Society for Clinical Investigation, which he served as president, on the subject "The Spirit of Science." In it he referred to the five objectives of the society, but the one given most prominence was "The diffusion of a scientific spirit among its members." In his remarks Lefkowitz contrasted the public perception of a scientist (dull, pedantic, aloof, devoid of emotion) with an artist (lively, colorful, high-spirited), although the genesis of scientific research is often similar to the work of an artist. It is a keenly felt sense of wonder and curiosity that translates into a genuine enthusiasm for even the faintest glimpse of understanding. A second aspect of the scientific spirit relates to the cultivation of the intuitive side of the scientist's nature; Lefkowitz had this to say: "When it comes to making creative scientific discoveries, imagination is perhaps more important than knowledge."

Lefkowitz was addressing a professional group, but much of what he said was directed to students. If interpreted literally, it might be shocking. He complains, for instance, that most of our institutions and granting agencies are skewed toward the rational ideal while the role of intuition is rarely even acknowledged. He contends that among the attitudes that encourage the recognition of serendipity and the flow of intuition are humor and playfulness: "A frequently exercised sense of humor favors the kind of wild, occasionally illogical or offbeat leaps that are part and parcel of the creative process."

His reference to creativity is thought provoking because so much is still to be learned about the process, including the concept that creativity can be taught. We should remind ourselves that a thorough understanding is not needed for a process to be useful. For example, fermentation goes back to biblical times, but centuries elapsed before enzymes were recognized as the responsible agents. Progress is made as the result of questions being asked, even ones that might seem to be irrelevant.

Stupid Questions Can Be Important

Fergus W. Campbell, professor of neurosensory physiology at the University of Cambridge, said this in response to my inquiry of how progress was made:

> In an experimental subject like mine, serendipity inevitably plays an important role. The experimenter asks Nature "Is my guess right or wrong?" If the guess turns out to be right, one moves to the next experiment to challenge it more critically. If it is wrong, one sits and raises another stupid question.

Along with his letter he enclosed a collection of sayings of Edwin Land, the inventor of the Polaroid Land Camera. Evidently Land's employees often quote their boss in the same reverential fashion that a Chinese worker might quote Chairman Mao. Campbell's favorite Land quote is "Every creative act is a sudden cessation of stupidity."

The Accidental Drop of a Tool

In his graduate studies Ross E. Davies, Scripps Institution of Oceanography at La Jolla, California, was observing wave actions in a flow tank when a fellow graduate student happened to drop a tool into the tank, thereby producing an unexpected wave motion. It took two years of study but eventually Davis was able to explain the motion and a new class of solitary waves was defined. Later, in recounting the story, Davis offered a definition: "Luck is really just an interpretation to occurrences which cannot be predicted."

The Necessity of Being Receptive

Leonard G. Goodwin, former director of science for the Zoological Society of London, feels that everyone needs a bit of luck to be at the right place at the right time. He also feels, however, that some of the things ascribed to "chance" are more likely to be the fitting together of events and concepts by minds that are receptive to them. This theme was repeated several times in the responses received from noted scientists.

The Inevitability of Chance

A letter from Daniel McKenzie, Department of Earth Sciences, University of Cambridge, indicated his preference for the term "luck" rather than "chance." Webster's *New Collegiate Dictionary* defines chance as "something that happens unpredictably without discernible intention or observable cause," while luck is "a force that brings good fortune or adversity." There can undoubtedly be cogent reasons for preferring one term over the other, but the key element seems to be the unpredictable nature of the event.

McKenzie raised an alternate question: "What parts of someone's career were *not* due to chance?" Since he had been a teenager, he wanted to do scientific research, but had no idea in what subject, where he should be, or how to set out on such a career. Fate evidently dealt kindly with him in his path toward membership in the prestigious Royal Society.

John M. Edmond, professor of geochemistry at the Massachusetts Institute of Technology, is a member of the British Royal Society and he points out that "luck," in the sense of recognizing and exploiting opportunities, is a learnable skill. Furthermore, he feels that teaching it forms the basis of all good educational programs. (I am forced to wonder, however, how Luck 101 might be regarded as an elective course.)

In response to my letter to him, Alec D. Bangham, former head of the biophysics unit at the Agricultural Research Council Institute of Animal Physiology, Babraham, Cambridge, said that "awareness" would be a more appropriate word to use.

For Michael Smith, director of the Biotechnology Laboratory of the Medical Research Council of Canada, unplanned circumstances led to his choice of a scientific field, and chance played a role in more than one of his accomplishments. Moreover, his experience indicated that it is unlikely that many scientists have had *planned* careers, and Smith is sure that chance discovery plays a role in many scientists' successes.

Hans Selye, the eminent endocrinologist, said: "The element of chance in basic research is overrated. Chance is a lady who smiles upon those few who know how to make her smile."

In the life of a researcher are many instances in which experiments fail, often through no fault in design. Things just don't work out as planned. Less frequent, but infinitely rewarding, are those positive results that elicit shouts of "Well, I'll be damned!" Hans Selye's comment is accurate because it is one thing to get a lucky break, and another thing to recognize it. Unfortunately, not all scientists bring an open mind to the laboratory bench.

Capitalizing on a Lucky Break

Heinrich Rohrer, winner of the Nobel Prize (with Gerd Binnig), for the development of the scanning tunneling microscope, made an apt comment in his response to me: "Getting the cards is one aspect, playing the hand the other." He ended with this commentary:

> When I started to think about new research directions for myself even earlier, I did not anticipate the scanning tunneling microscope. So I was lucky. And in this sense, you constantly are confronted with circumstances you did not plan but you can grasp to make the best out of it. To put it differently, most people have quite a lot of luck along the way, but only few perceive it.

A Prospector in the Genetic Code

Carl R. Woese, professor of biology at the University of Illinois at Urbana, became interested in genetics in the early 1960s when he was with the General Electric Company at their main research laboratory in Schenectady, New York. It was while he was waiting for equipment to be delivered that evolutionary biology occupied his thoughts. At that time it was an accepted fact that cells can be classified as either eukaryotes (organisms having a nucleus) or prokaryotes (organisms without a nucleus). Protein synthesis occurs in what are called ribosomes, and Woese wondered why eukaryotic and prokaryotic ribosomes were so different. To complicate the problem, within the prokaryotes there are distinct groups of organisms that seem to be no more related to one another than they are to eukaryotes. His work involved comparing the sequences in ribosomal DNA and led him to conclude that there was a third branch of life, called the archaea. A definition of archaea would set them quite apart from, on one hand all plants, and animals on the other. They are a group of organisms that have sequences of ribonucleic acids, coenzymes, and cell walls that differ from all other organisms, and they are considered to be an ancient form of life that evolved separately from bacteria and blue-green algae.

Woese's response to my letter stated, in part:

> The situation feels something like prospecting for silver. To find a strike is always a matter of chance. However, good prospectors have a far better chance of finding one than poor prospectors. In this case, too, it was like prospecting for silver and chancing upon gold.

Striking Pay Dirt at NIH

Very similar to Woese's remarks were those of Robert H. Purcell, head of the Hepatitis Viruses Section at the National Institutes of Health in Bethesda, Maryland. His coming to NIH and then entering the field of hepatitis research were not planned. Furthermore, Purcell admits that it was a chance decision to accept Mario Rizzetto into his laboratory. Their collaboration on the discovery of the hepatitis gamma-virus led to studies that revealed the existence of blood-borne non-A, non-B hepatitis virus, and the enterically transmitted non-A, non-B hepatitis virus. He commented:

> I often liken scientific research to prospecting for minerals (something I have done as a hobby in the past) because both require elements of chance and very hard work.

An Opposing View

In the interest of balance I nod to someone who has written critically about the role of serendipity, that being the renowned Philip Abelson. Readers of *Science* know Abelson as its editor for many years, but he had made his reputation long before assuming that exalted position. It was Abelson, when he was at the Naval Research Laboratory in the early 1940s, who urged strongly that atomic energy be used to power the Navy's submarines. Regarding serendipity, however, Abelson wrote a negative editorial on serendipity which contained these passages: [1]

> Occasionally a chance observation has led to unexpected enlightenment. In general, however, progress has come because experimenters were seeking it.

> and

> In general, the research worker gets no more from his experiments than he puts in by way of thought, preparation, performance, and analysis. Serendipity is a bonus to the perceptive, prepared scientist, not a substitute for hard work.

Mentors Can Affect Career Choices

In today's world there is much discussion of role models, mostly as the subject applies to professional athletes. It is a different matter with scientists operating on a much different level, but a professor's influence on a student can be enormous.

John Christman's encyclopedic knowledge of serendipity caused him to wonder about the influence of professors on their students. In his study of teacher/student relationships, he was led to conclude that those teachers who were aware of the potential importance of serendipity in their research inculcated the concept in their students, making them more aware of the possibility that unplanned circumstances might be important factors in their research.

Alec D. Bangham, mentioned earlier, would have reason to question the positive role of a mentor. He was elected a Fellow of the Royal Society in recognition of "work on the structure of phospholipids in aqueous media and especially for developing the liposome as a model for cell membranes." He wonders what might have been the course of history if he had followed the advice of his boss, who sent him a memo requesting that he work on "real membranes."

That brings to mind an observation made often while doing the research for this book, namely that scientists can succeed despite the inadequacies of their supervisors. Sometimes, as noted earlier by Fergus Campbell, teachers might credit students for asking them questions that force them to examine critically the positions they have taken. Sometimes this forces a change in their thinking.

A Final Thought

Research will continue to be based on the scientific method that has served so well in all of the scientific disciplines. It is probably inevitable, however, that serendipity will continue to play an important role in discoveries, just as it has in the past. Scientists should be sufficiently humble to realize that their well-thought-out hypotheses might require change when confronted with facts that had not been even imagined.

REFERENCES

1. Abelson, Philip. 1963. Editorial. *Science.* 140:No. 3572.

Index

A

Abelson, Philip 213
Adams, Roger 31
Adler, Erich 97
Adrian, Edgar 158
Alfven, Hannes 126, 161
Allison, Anthony 145
Altounyan, Roger 161
Alvarez, Julian 42
Alvarez, Luis 158, 159
Anderson, Carl 142, 143
Andrew, Captain Frank 200
Andrewes, Sir Christopher 84
Askonas, Brigitte A. x, 168
Atkins diet 18, 19
Avery 69
Axelrod, Julius 156, 174

B

Bangham, Alec D. 211, 213
Baranyi, Robert 148
Barcroft, J. 35
Barger, Bill xi, 54
Barger, George 86
Baur, Erwin 170
Bazett, Cuthbert 178
Beaudette, Fred 71, 118
Bedson, Samuel 89
Behre, C. H., Jr. 128
Bell, Jocelyn 132
Bennich, Hans 12
Bernheim, F. 75
Beroza, Morton 122
Best, C. H. 81
Bethe, Hans 132
Bigelow, Henry 172
Binnig, Gerd 211
Blackett, P. M. S. 136
Blanchard, Duncan 52, 55, 59, 165, 173
Blumberg, Baruch 145
Bodmer, Walter x, 177
Boer, H. 29
Borel, Jean-Francois 87
Bothun, Gregory D. 130
Boyer, Paul 101
Boyum, Arne 106
Bridgman, Percy 191
Brodie, Steve 156
Brown, Donald 174
Brown, Frank 108
Brown, Hanbury 129
Brown, Herbert 152, 153
Bugie, E. 72
Burdsall, 99
Burnet, Sir Macfarlane 176
Burnstock, Geoffrey 104
Bush, William 88

C

Cahill, George F. 15
Calne, Donald B. 159
Calvin, Melvin 175
Campbell, Fergus W. x, 209
Campbell, Harold 83
Campbell-Renton, Margaret 62

Carhart, Homer x
Carliner, P. E. 91
Carlson, Ed 82
Carson, Rachel 30
Chain, Ernst Boris 60
Chandrasekhar 132
Chibnall, Charles 175
Christman, John ix, xiv, 13, 28, 30, 43, 162, 213
Christy, James 134
Chu, Joseph 13
Cleland, Wallace 85
Coates, Admiral 200
Cocking, Edward C. x, 105
Cohen, Stanley 150, 151
Cole, Warren H. 157
Conrad, Joseph 50
Costerton, J. William 10
Cowdry, E. V. 39
Cowling, Ellis 97
Cox, James L. 156
Cravin, Lawrence 84
Crowe, John H. x, 46
Curl, Robert 34
Curtis, Lt. Denny 201

D

Dale, Henry H. 86
Dalgarno, Alexander 167
Davenport, Horace 178
Davis, Bernard D. 162
Dawson, M. H. 63
Dexter, T. Michael 158
Dripps, Robert D. 92
Dubos, Rene 69, 71

E

Edmond, John M. x, 210
Eglinton, Geoffrey x, 175
Ehret, Gunther 121
Ehrlich, Paul 70
Eijkman, Christiaan 151
Eriksson, Karl-Erik 99

F

Feibelman, Walter 133
Felbeck, Horst 167
Feldman, W. H. 73
Finland, Maxwell 88
Fitzsimmons, Vincent 192
Flemer, Dave 3
Fleming, Alexander 16, 61, 67, 70, 142
Florey, Howard 65
Fowler 132
Fox, Harry 192
Fox, Herbert 91
Fraenkel, G. S. 20
Frankel, O. H. x, 170
Fraser-Smith, Antony 135
Friedman, Herbert 171, 186
Furchgott, Robert 39
Fyffe, L. R. 175

G

Gajdusek, D. Carleton 145
Garden, N. P. 31
Garrett, Bill 54
Gay, L. N. 91
Gell-Mann, Murray 4
Glaser, Donald 142, 143
Goodwin, Leonard G. 210
Gortner, Ross 82
Graham, Evarts A. 157
Green, Howard 122
Griffith, James 171
Gruberg, Edward R. 18
Guillette, Louis J. 117

H

Hagen, John P. 198
Halle, Morris x, 170
Hamburger, Viktor 150
Hare, Ronald 62, 65
Harrison, G. G. 117
Hastings, Baird 102
Hastings, Woodland x
Hauptman, Herbert x, 171

Hay, Allan S. x, 168
Heatley, Norman 64
Hess, Viktor 143, 144
Hileman, Bette 117
Hill, R. T. 90
Hinshaw, H. C. 73
Hirschhorn, Kurt 17
Hitchings, George 182
Hitchner, S. B. 118
Hochachka, Peter 167
Hodgkins, Alan Lloyd 175
Holliday, Robin x, 169
Holmes, Robert 120
Holt, E. 81
Hopkins, F. Gowland 35
Howard, Lt. John 201
Howell, William 80
Hubel, David 149
Huebner, C. F. 83
Huffman, Donald 33
Hughes, John 19
Hukovic, Seid 105
Hutchison, Robert 88

I

Ignarro, Louis 39
Impey, Christopher D. 130
Irving, Edward (Ted) 136
Isaacs, Aleck 84
Iselin, Columbus 172

J

Jackson, Tommy 43
Jakobson, Roman 170
Jansky, Karl 125
Jeffay, Henry 42
Jensen, James 32
Jobe, Frank 2
Johansson, Gunnar 12
Johnson, C. G. 177
Jones, A. 131
Jorpes, Erik 81
Juhlin, Lennart 12

K

Kamiokande 131
Kanigel, Robert 23, 174
Kappas, Attallah 18
Karle, Jerome 171, 205
Kauffman, George 74, 89
Kebabian, John W. 159
Keilin, D. 35
Kelman, Arthur 97
Kessler, Steven 12
King, Peter 184, 186, 197
Kirk, T. Kent x, 96
Kleiber, Max 31
Klemme, Dorothea 13
Knudson, Alfred G. x, 166
Kolata, Gina 39
Kolbe 79
Kornberg, Arthur 60, 151
Kosterlitz, Hans 19
Kroto, Harold 33, 34
Kuc, Joseph 174
Kuffler 38, 149
Kuffler, S. W. 38
Kunin, Calvin M. 88, 94

L

Land, Edwin 209
Langmuir, Irving 51, 101
Latham, David W. 130
LaTouche, C. J. 62
Lauteman 79
Laws, Richard M. 14
Lederberg, Joshua 162, 176
Lederman, Leon 141
Lefkowitz, Robert J. x, 208
Lehmann, Jorgen 75
Lemberg, Max Rudolf 36
Leonard, John M. x, xi, 22, 183, 193
Leopold, Luna B. 166
Levi-Montalcini, Rita 150
Levy, David 125
Light, Alan 120
Lindenmann, Jean 84

Link, Karl Paul 82
Lipsky, S. R. 30
Lockhart, Luther 187
Long, C. N. H. 178
Los, Marinus 182
Lovelock, James 28
Lowell, J. R. 124
Lowell, Percival 134
Lucas, Keith 175
Lundquist, Knut 100
Luria, Salvador 162

M

Macfarlane, Gwyn 62
Maeda, K. 135
Malin, David F. 130
Martin, A. J. P. 28, 30, 142
Maurois, Andre 63
Mayer, Ben 133
Mazeh, Tsevi 130
McKenzie, Daniel x, 210
McLean, Franklin 102
McLean, Jay 80, 81
McNaught, R. H. 131
Medawar, Peter 181, 182
Midgely, Thomas 5
Mitchelson, F. 104, 114
Moncada, Salvador 40
Mooney, J. B. 201
Morse, Jared 102
Moyer, Andrew Jackson 65
Murad, Ferid 39, 40

N

Nalbandov, A. V. 90
Narins, Peter 121
Nathanson, James A. 116
Neihof, Rex xi, 14
Newburgh, L. H. 17
Newman, Mel 175
Nobes, David C. 3
Nordling, Carl 42
Nossal, Sir Gustav 176, 182

O

Ogston, A. G. 145
Olson, C. Marcus 36

P

Page, Robert 184
Palevitz, P. A. 41
Papanicolau, George N. 104
Pasternak, Gavril 23
Paulley, J. W. 160
Peltier, Athanase 56
Pert, Candace 10, 23
Peterson, Ivars 130
Pfuetze, Karl 74
Phillips, M. M. 131
Plunkett, Roy ix, 5, 7, 192
Polge 163
Purcell, Robert H. x, 212

Q

Quick, A. J. 83

R

Rajfer, Jacob 41
Rand, M. J. 104
Raper, Kenneth 65
Raymond, Stephen A. 18
Rhines, Chester 70
Richet, Charles 119, 142
Richter, Curt 91
Roberts, Eugene x, 37, 162
Robinson, Sir Robert 175
Roedder, Edwin x, 128
Rohrer, Heinrich 211
Rowan, Tom 39
Roy, Rustum ix, xiv, 162
Rydon, H. Norman 45

S

Sandin, Reuben Benjamin 168
Sargent, Marston 9
Schaefer, B. 135
Schaefer, Vincent 51

Schairer, J. Frank 128, 137
Schatz, Albert 72
Schemroth, Leo 160
Schmidt, Maarten 132
Schulman, J. H. 101
Selye, Hans 211
Shelton, Ian 131
Siegbahn, Kai 42
Skellett, Melvin 126
Skoog, Folke 8
Skou, Jens 100
Smalley, Richard 33, 34
Smith, Audrey 163
Smith, Michael 211
Smith, P. W. G. 45
Smoyer, Carol 118
Snowman, Adele 23
Snyder, Sol 24
Somero, George N. x, 166
Southwood, Sir Richard x, 177
Southworth, George 126
Stierle, Andrea and Donald 8
Stockard, Charles R. 103
Stow, Richard W. 9
Stroh, Michael 40
Sweeney, Beatrice 8
Synge, J. P. 28

T

Tabor, David 16
Tautz, Jurgen 121
Taylor, Albert Hoyt 11
Terenius, Lars 10
Thompson, Neil E. S. 34
Tiselius, Arne 156
Tombaugh, Clyde 134
Trubey, Robert 43
Tsipursky, Semeon J. 34
Tulving, Endel x, 9
Turro, Nicholas 35
Twiss, Richard 129

V

Vine, Allan 112
Visnich, Sam 146
Vogt 176
Volta, Alessandro 56

W

Waksman, Selman 68
Waldrop, M. Mitchell 129
Walker, Donald x, 169
Walker, John E. 101
Wallerstein, George 133
Ward, James T. 47
Wasmer, Otto 96
Watkins, George D. x, 172
Watson 129, 169
Watson-Watt, Sir Robert 129
Werner, Henry 44
Whitaker, D. B. 106
Whitby, Lionel 90
Wiggers, Carl J. 178
Williams, Kingsley xi, 184
Wilson, C. T. R. 142, 143
Wilson, Grove 173
Wishart 177
Wittig, George 27
Wolman, M. Gordon x, 166
Woodcock, Alfred 172
Woods, D. D. 75
Woosley, S. E. 131
Work, Thomas 168
Wright, Irving 81
Wright, Sir Almoth 70

Y

Yale, Harry L. 75
Young, Leo 11

Z

Zasloff, Michael 116
Zinder, Norman 162
Zisman, William 183, 191

If your local book store does not have this book:

Send an e-mail to info@aidthroughtrade

or

Call 410-266-8857

or

On the Internet, bring up www.iUniverse.com**, then click on Browse the Books, and then fill in the book title, and Search**

or

Call 1-800-AUTHORS

978-0-595-36551-7
0-595-36551-5

3034726

Made in the USA